DIGITAL SIGNAL PROCESSING EXPERIMENTS

using a personal computer with software provided

Alan Kamas

Edward A. Lee

University of California at Berkeley

PRENTICE HALL, Englewood Cliffs, New Jersey 07632

Library of Congress Cataloging-in-Publication Data

Kamas, Alan
Digital signal processing experiments using a personal computer with software provided.

Bibliography: p.
1. Signal processing—Digital techniques—Experiments—Data processing. I. Lee, Edward A. II. Title.
TK5102.5.K315 1989 621.38'043 88-25507
ISBN 0-13-212853-5

Editorial/production supervision: *Mary Rottino*
Manufacturing buyer: *Mary Ann Gloriande*

A Division of Simon & Schuster
Englewood Cliffs, New Jersey 07632

The publisher offers discounts on this book when ordered in bulk quantities. For more information, write:

Special Sales/College Marketing
Prentice-Hall, Inc.
College Technical and Reference Division
Englewood Cliffs, New Jersey 07632

Printed in the United States of America
10 9 8 7 6 5 4 3

ISBN 0-13-212853-5

Prentice-Hall International (UK) Limited, *London*
Prentice-Hall of Australia Pty. Limited, *Sydney*
Prentice-Hall Canada Inc., *Toronto*
Prentice-Hall Hispanoamericana, S.A., *Mexico*
Prentice-Hall of India Private Limited, *New Delhi*
Prentice-Hall of Japan, Inc., *Tokyo*
Simon & Schuster Asia Pte. Ltd., *Singapore*
Editora Prentice-Hall do Brasil, Ltda., *Rio de Janeiro*

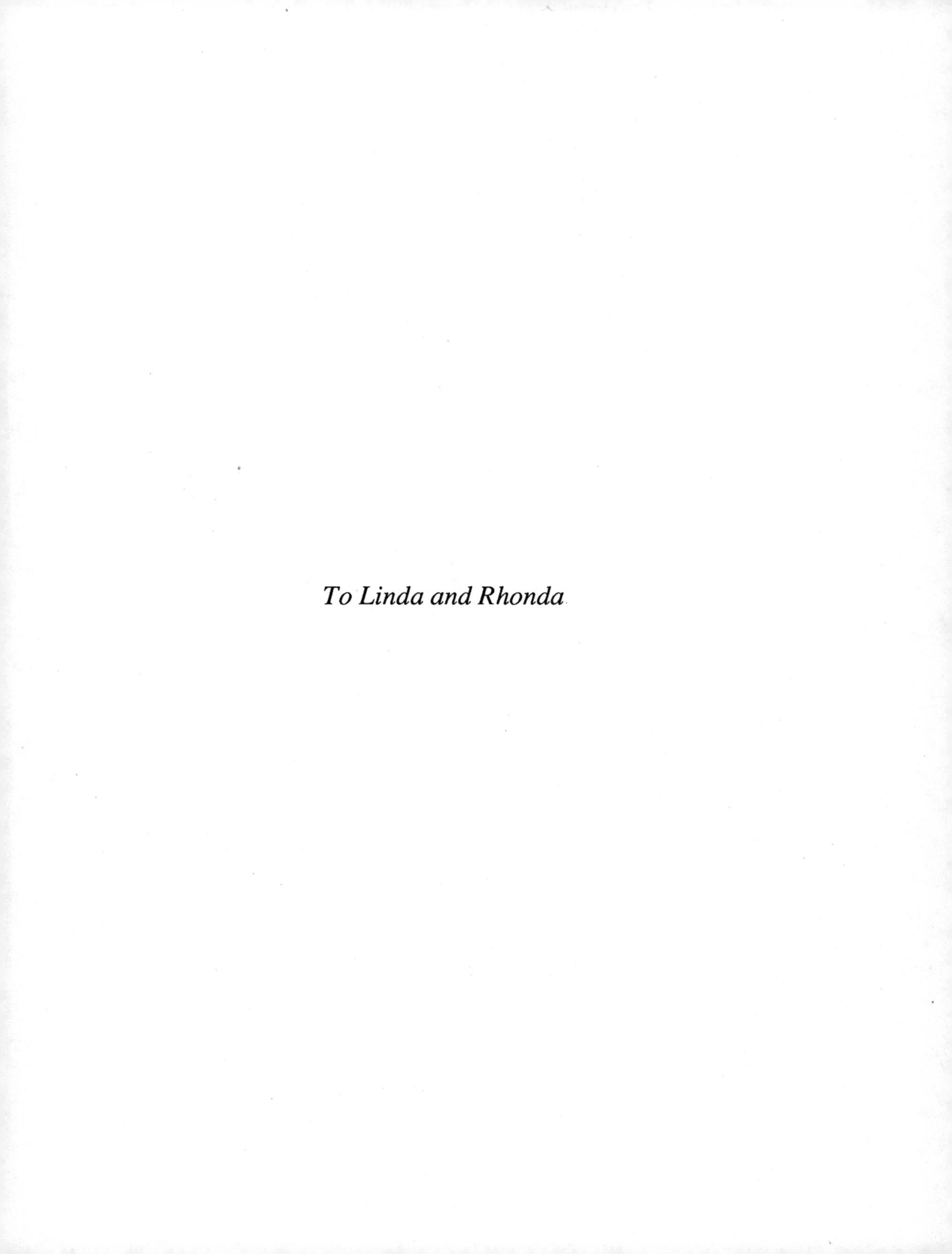

To Linda and Rhonda

CONTENTS

INTRODUCTION

Intuition about digital signal processing is difficult to solidify in the abstract, without practical experience. But gaining practical experience usually involves tedious hardware and software projects where most of the effort has little to do with signal processing and much to do with computer architecture, microcoding, or circuit design. Fortunately, recent software advances have provided an alternative. This lab manual is based on a program called DSPlay[1], developed by the Burr-Brown Corporation, and generously donated to this project. It permits the student to experiment with sophisticated digital signal processing systems without having to learn a great deal about irrelevant mechanics.

DSPlay is interactive, easy to learn, and encourages exploration. DSP systems are built graphically as block diagrams using higher level signal processing functions such as digital filters and FFTs. The block diagrams are called *flowgrams*. A reasonably complete set of primitive functions (signal generators, adders, multipliers) permit a surprising variety of signal processing concepts to be demonstrated. The student gets immediate feedback about the chosen configuration, and therefore can reason about errors (or inventions). Indeed, some of the exercises in this manual deliberately yield surprising results which are easily explained in retrospect, but rarely anticipated by the student.

Although the theory developed in a typical course on digital signal processing is mature, there are numerous concepts that are difficult to convey with mathematics alone. Examples help, but for many concepts, examples involve too much computation to be worked out by hand in the lecture or

[1] DSPlay is a trademark of Burr-Brown Corporation.

in the homework. Some such concepts are filter design, circular convolution, graphical interpretation of pole-zero plots, linear-phase filtering, the DFT as compared to the discrete-time Fourier transform, overlap and add, overlap and save, and quantization effects. The emphasis of the laboratory exercises in this manual is on supplementing abstract study with concrete demonstrations of such concepts. It is not the intention to replace abstract study, and concepts that are better explained with mathematics do not appear explicitly in the lab manual.

The eleven exercises are carefully designed to be done in order, with the possibility of skipping lab 6 and lab 11. The ordering is important because the requisite skill with the DSPlay software is developed incrementally. Furthermore, signal processing concepts are developed incrementally, and later labs rely on intuition developed in earlier labs. Some of the labs use functions that are not expected to have been covered in the course lectures by the time they are used. For example, lab exercise 3 uses the FFT block to generate frequency domain plots, although the FFT is likely to be covered in the course much later. The student is asked to assume that the FFT block ideally computes the Fourier transform. The alert student will note artifacts in the display which should arouse curiosity. Those artifacts are explained later as side effects of the FFT.

This lab manual is intended to be self-contained, in the sense that a DSPlay manual is not need. All required DSPlay functions are explained in the lab exercise in which they first appear, and to facilitate learning most are explained more than once. They are also summarized in the appendix. The software fortunately has extensive "help" menus, so memorization of arcane commands is not required. The early exercises begin with tutorial "example procedures" which introduce the student to all the mechanics required to solve the problems that follow. The later exercises do not require any new mechanics, and consequently proceed directly to the problems.

1. OVERVIEW OF THE LABS

There are eleven laboratory exercises. They begin with elementary signal processing concepts such as convolution and progress to advanced issues such as quantization effects.

Lab 1: *Signal Generation and Convolution*

The example procedures introduce the student to DSPlay by explaining every step required to generate and convolve two square pulses, display the results, and interpret the plots. The first problem is to construct a minor variation on this simple system. The second problem is to use the primitives in DSPlay to synthesize an exponentially decaying signal. This lab deliberately requires the student to do it the hard way, with a much easier way developed in lab 2.

Lab 2: *The Z Transform and System Functions*

Students use the software to find the inverse Z transforms of rational polynomials in z. The IIR filter block in DSPlay is used even though IIR filter structures will not have been explained in the course at this time. It is used as a general implementation of a rational system function. A much simpler solution to the problem of generating an exponentially decaying sequence is now possible by computing the inverse Z transform of a first-order, one-pole Z transform.

Lab 3: *Pole/Zero Plots and Frequency Response*

This lab explores the relationship between rational Z transforms, their pole/zero plots, and their frequency responses. The FFT block is used to estimate frequency response even though the student is not expected to understand the mechanics of the FFT yet. The student gets exposure to resonators, all-pass filters, and comb filters.

Lab 4: *Linear Phase Filtering*

This lab introduces linear-phase FIR filtering and demonstrates the difference between even- and odd-symmetric filters. The lab also looks at group delay and group symmetry. In addition to FIR filters, the lab introduces non-causal filtering using signal reversal to accomplish linear-phase filtering.

Lab 5: *Amplitude Modulation*

As an application of filter design, the students are asked to generate an interesting signal, modulate it, filter out one of the sidebands, demodulate it, and low pass filter the result. If the demodulation frequency is different from the modulation frequency, then a spectral shift results. This effect is demonstrated.

Lab 6: *Spectral Shifter*

This exercise requires the use of complex-valued time-domain signals. If these are not being covered in the course, this lab should be skipped. The objective is to construct a spectral shifter (like that in lab 5) using complex-valued signals. This method admits a much lower sampling rate than the method used in lab 5. Students learn the practical value of complex-valued signals, analytic signals, and Hilbert filters.

Lab 7: *The DFT, FFT, and Circular Convolution*

The DFT and circular convolution are used, explained, and explored. The use of zero-padding and the impact of DFT length are emphasized, since these ideas are crucial to proper use of the tools.

Lab 8: *IIR Filter Design*

This lab demonstrates both the impulse-invariant transformation and the bilinear transformation for IIR filter design from continuous-time filters. This gives the student practical experience with computer aided filter design tools.

Lab 9: *FIR Filter Design via Windowing*

Students design and test FIR low-pass and high-pass filters using rectangular and Hamming windows. The linear phase response of the resulting designs is demonstrated by showing how symmetry of an input pulse results in symmetry of the output pulse.

Lab 10: *Frequency Sampling and Equiripple FIR Designs*

The frequency sampling method for the easy design of filters with almost any desired frequency response is demonstrated. Also, the Parks-McClellan algorithm is used to generate multi-band optimal equiripple filters.

Lab 11: *Quantization Effects*

This lab investigates the effects of coefficient quantization on filter response. Narrow band second order IIR filters are examined. Multirate filtering is used to implement very narrow band quantized IIR filters. The effects of coefficient quantization on the frequency response of FIR filters are also demonstrated.

2. SUPPORTED TEXTS

The lab exercises are explicitly keyed to the following texts:

A. V. Oppenheim and R. W. Schafer, *Discrete-Time Signal Processing* [Prentice-Hall, Englewood Cliffs, NJ 1988].

A. V. Oppenheim and R. W. Schafer, *Digital Signal Processing* [Prentice-Hall, Englewood Cliffs, NJ 1975].

L. B. Jackson, *Digital Filters and Signal Processing*, second edition [Kluwer Academic Press, Boston, MA 1988].

J. G. Proakis and D. G. Manolakis, *Introduction to Digital Signal Processing* [Macmillan, New York, 1988].

R. D. Strum and D. E. Kirk, *Discrete Systems and Digital Signal Processing* [Addison-Wesley, Reading, MA, 1988].

R. A. Roberts and C. T. Mullis, *Digital Signal Processing* [Addison-Wesley, Reading, MA, 1987].

D. J. DeFatta, J. G. Lucas, and W. S. Hodgkiss, *Digital Signal Processing: A System Design Approach* [John Wiley & Sons, New York, 1988].

R. Kuc, *Introduction to Digital Signal Processing* [McGraw-Hill Book Co., New York, 1988].

L. R. Rabiner and B. Gold, *Theory and Application of Digital Signal Processing* [Prentice-Hall, Englewood Cliffs, NJ, 1975].

A. V. Oppenheim and A. S. Willsky with I. T. Young, *Signals and Systems* [Prentice-Hall, Englewood Cliffs, NJ, 1983]. (Referred to in some laboratory exercises as a background reference)

In addition to specific keying, each of the laboratory exercises contains an outline of the background material the student is assumed to know. By using these outlines, the labs may be easily integrated into a course that is not based on one of texts above.

3. NOTATION

The notation adopted is identical to that in Jackson, and similar to that of the other texts. We use T to denote the time between samples. The frequency response of a system with impulse response $h(n)$ is written $H'(\omega)$ and is the same as the discrete-time Fourier transform of $h(n)$.

4. HARDWARE AND SOFTWARE REQUIREMENTS

The only hardware required for the exercises in this manual is an IBM PC, XT, AT, or compatible with at least 512K of RAM and a color graphics adapter (CGA) or enhanced graphics adapter (EGA). An IBM graphics printer or compatible is desirable, but not absolutely necessary. The exercises have been designed to require a minimal number of printouts, and if a printer is not available, the screen display can be easily sketched by the student where required.

The only software required, included in a disk at the back of this manual, is a simplified version of DSPlay from Burr-Brown. In order to reduce the program to one disk, certain important capabilities of DSPlay have been omitted. For example, the software provided here makes no use of coprocessors, and consequently runs considerably slower than the complete version available from Burr-Brown. The maximum buffer size (the maximum signal length) in the version of DSPlay provided is smaller than the complete version. Also, the interface to A/D and D/A hardware has been omitted, as has the ability for the user to define new blocks. Finally, the code generation capability for real-time DSPs has also been left out. The complete version of the software may be obtained from Burr-Brown, and site licenses are available, with special treatment given to universities. Contact the Burr-Brown Corporation, P.O. Box 11400, Tuscon, AZ 85734, USA.

5. NOTES TO THE INSTRUCTOR

The intent of the authors is that this lab manual be used in an undergraduate or first-year graduate digital signal processing course as a supplement to problem sets. It is used at U. C. Berkeley in a senior level course which has three hours of lecture per week for a 15 week semester. Students are expected to spend an average of three hours per week in the lab.

In order to minimize the cost of instituting a lab based on this manual, we have chosen to omit exercises involving acquisition or production of real-time signals. Although such exercises are possible with the complete version of the DSPlay software, they have hardware requirements that may be costly and difficult to orchestrate for large classes.

Unfortunately, any software that permits application descriptions at this high a level will have its limitations. The version of DSPlay provided here has limited ability to handle flowgrams with feedback. Because of the limitations, we have chosen not to use the feedback capability in the software. Surprisingly, this means remarkably little compromise of content in the exercises. Feedback can always be implemented within a block in the flowgram, and the IIR filter block provides a general framework for building linear systems with feedback. The only loss is the explicit illustration of the feedback in the flowgram, which may have served to aid intuition.

6. ACKNOWLEDGEMENTS

Our deepest appreciation goes to the Burr Brown Corporation, who has donated the software to this project. In particular, we are indebted to Ron Walk and George Radda of Burr Brown, without whose enthusiasm and cooperation this project would have been much more difficult. Equally important was the support of the Council on Educational Development of the University of California at Berkeley, who provided a grant to support this project. Further support was also provided by the College of Engineering and the Electrical Engineering and Computer Science Department at UC Berkeley. We are also indebted to the IBM corporation, whose generous gift of PC ATs made it possible for us to implement and test the lab. A number of individuals contributed helpful suggestions, particularly Dev Chen, Eric Cox, Darryl Sale, and Andrew Sekey. Finally, our thanks extend to the many students of EECS 123 at Berkeley, who endured early versions of this lab and offered constructive criticism.

Laboratory Exercise 1

Signal Generation & Convolution

Purpose:

The example procedures introduce the student to DSPlay by explaining every step required to generate and convolve two square pulses, display the results, and interpret the plots. The first problem is to construct a minor variation on this simple system. The second problem is to use the primitives in DSPlay to synthesize an exponentially decaying signal.

Background Material:

This lab only requires an understanding of the basics of convolution. Convolution and discrete time systems are covered in:

Oppenheim and Schafer [1988]: Sections 2.1 - 2.4

Oppenheim and Schafer [1975]: Chapter 1

Jackson: Chapter 2

Proakis and Manolakis: Sections 2.1 - 2.3

Strum and Kirk: Sections 2.0 - 2.2, 2.4, and 3.4

Roberts and Mullis: Sections 2.1 - 2.4

DeFatta, Lucas, and Hodgkiss: Sections 2.0 - 2.2

Kuc: Sections 2.1 - 2.5

Rabiner and Gold: Sections 2.1 - 2.5

Oppenheim and Willsky with Young: Chapter 3

The procedures below are intended to be self-contained. Further information, however, on all DSPlay functions and commands may be found in the appendix.

Example Procedures:

Perhaps the best way to learn a new software package is to simply sit down and use it. Thus, it is highly recommended that the student work through the following step-by-step procedures before starting the problems.

Convolving two square pulses

Create 3 Blocks

1) Start the DSPlay program. In most configurations, it is adequate to power-up the PC and type "dsplay" followed by <Enter>. If this fails, follow the instructions of the instructor. You should now have an almost empty screen with an empty function block on the far left.

   At the top of the screen the available commands are summarized. Typing the capitalized letter of any of these commands will execute them. Try quitting and restarting the program. At the bottom of the screen, more commands are summarized. These are activated by special function keys on the keyboard. The most useful one is <F1>, the help key. Press it. The bottom of the screen will be updated to tell you what your current options are. The <Esc> key will escape from help mode. As a general principle, the <Esc> key gets you out of whatever you've gotten into.

   The program starts with the cursor at the bottom of the screen pointing to a prompt: "FLOWGRAM:". The system is said to be in "flowgram mode"; commands entered in flowgram mode operate on the entire flowgram. To operate on parts of the flowgram, or to create a flowgram, press the <up arrow> key, "return to block". The system is now said to be in "block mode"; the cursor, indicated by a flashing box, selects the block that commands will affect. Note that a new set of commands is summarized at the bottom of the screen.

2) Using the <Right Arrow> key (on the numeric key pad of some machines), move the cursor to the right of the existing block. Press the <Insert> key (this key is labeled <Ins> and may also be on the numeric key pad. A new block will appear at the cursor location.

3) Now build a new block just below the first block. Your three new blocks should be in the same positions as the three completed blocks in figure 1.

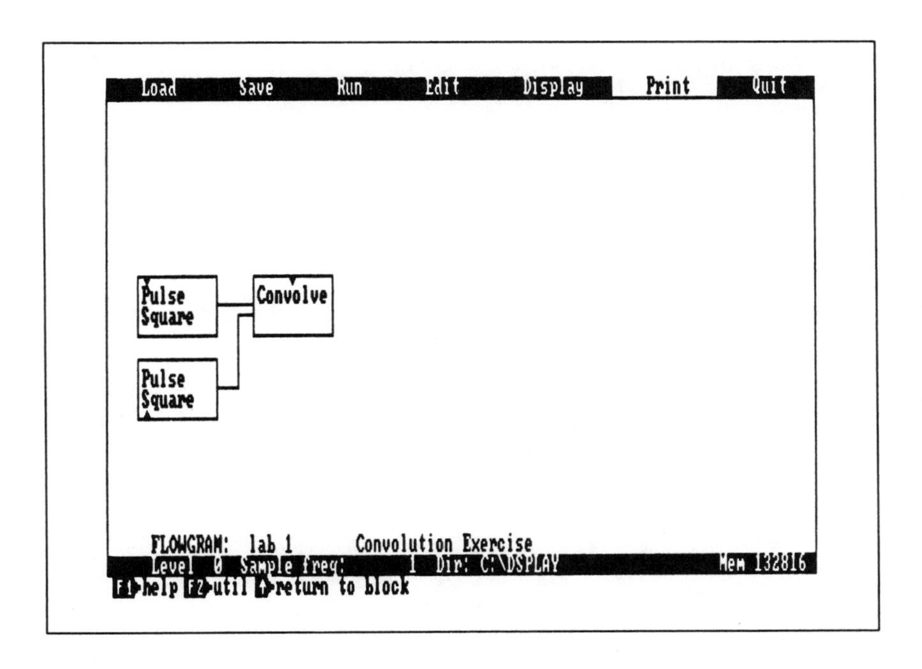

Figure 1. The finished flowgram from the example procedure.

Define the Blocks (2 pulse blocks and a convolver)

First Pulse Input Block:

1) The cursor should still be on the lower left block. Press <E> to enter the edit mode. A menu will pop down from the top of the screen.

2) Using the arrow keys you can select an entry in the menu, then pressing <Enter> will activate the selection. Alternatively, you can type the capitalized letter of the command in the menu. Type <P> to select the "Parameter" function. Note: with the cursor on a block, pressing <Enter> is a shortcut for <E> followed by <P>.

3) The "Parameter" function is used to select the parameters of a block. A long menu with the header "Functions" should be showing. Again, entries can be selected with <arrow> <Enter> or by typing the capitalized character of the entry. Press <I> to open the menu of Input functions.

4) Input functions are signal generators. The list of available input functions should be showing. Press <P> to get a pulse. The pulse block can be used to get a piecewise linear pulse. We will use it get a square pulse.

5) A new type of menu appears that looks like a form with entries to be filled in. The first entry is the name you will assign to this particular input block. Type for example "Square" <Enter>.

6) The pulse generator can produce either a periodic pulse stream or a single pulse. Type N<Enter> to request a single pulse.

7) A single pulse obviously does not have a period but the program requires that one be entered anyway. The period should be larger than (rise time) + (pulse width) + (fall time). Enter a large number; 1000 should work well.

8) Enter 5 for the pulse width. The pulse width is given in *seconds*. The system defaults to sampling at 1 Hz, so the pulse width will be 5 samples long.

9) Enter 0 for the rise time. The rise time here is the number of *seconds* (not samples) needed to linearly rise from zero to the amplitude of the pulse.

10) Enter 0 for the fall time. The fall time is the number of *seconds* required to linearly fall from the amplitude of the pulse to zero.

11) Enter 1 for the amplitude.

12) Press <Enter> to leave the offset voltage at zero and to leave the Pulse parameter menu. You have now finished building your first pulse block. It is labeled on the screen with the block type "Pulse" and the name you gave it "square".

Second Pulse Input Block:

1) The cursor should still be over the newly created pulse block. Press <E> and then <T> to tag this block for copying. The tagged block has its name displayed with reverse video.

2) Now move the cursor to the empty block above. Press <E> <C> to copy the tagged block into this empty block.

3) The reverse video of the first pulse block may be removed by "Untagging" the block. Move the cursor to the first block, use the <E>dit menu, and <U>ntag the block.

Convolution Block:

1) Move the cursor to the right hand block.

2) Press <Enter>. (This is the same as pressing <E> and then <P>.)

3) Select <F> for the filter menu.

4) Press <C> for the Convolver parameter menu. A convolver is a rather general form of a digital filter.

5) Enter a name for your convolver.

6) Type 32<Enter> for the buffer length. (See the discussion about buffer lengths in the appendix). You should now be back at the function block screen.

Connecting the blocks

1) The cursor should be over the Convolve Block. Now press the <Insert> key. You are now in the "line insert" mode. The cursor will now be a small dark dot over the convolver's upper input notch.

2) Press the <Left Arrow> key. This will cause a line to appear. Keep pressing the <Left Arrow> key until the line reaches the center right hand edge of the upper Pulse block. A connection now exists between the Convolve block and the upper Pulse block.

3) Press <Return> to signal to the program that the line is complete. The cursor will now be back over the Convolve block.

4) Press the <Insert> key again. Use the <Down Arrow> key to choose the Convolver's second input. The dot should now be over the second input notch.

5) Now use the <Left Arrow> key and the <Down Arrow> key to connect the second input of the Convolver to the output point of the lower Pulse block. Note: the output point of a block is always located in the center of it's right hand side.

6) Now press <Return> to complete the line. If all has gone well, your screen should now look like figure 1.

Marking the Blocks for Window Display

1) Move the cursor to the upper Pulse Block.

2) Press the <M> key and enter <1>. Now, whenever the flowgram is run, the output of this block will appear in window number one. The small dark triangle in the upper left hand corner of this block indicates window one.

3) Now move the cursor to the lower Pulse Block.

4) Press <M> and enter <2> for this block's window ID number. Note that the small triangle in this block is in the lower left hand corner.

5) Finally, move the cursor to the Convolve Block. Press <M> and enter <3> for the Window ID.

Naming and Running the Flowgram

1) Press the <Home> key (it is on the numeric key pad) to get into flowgram mode. When in this mode, operations affect the entire flowgram.

2) Press <E> to open the flowgram parameter menu.

3) Enter a name for your flowgram ("Lab 1" might be a good name).

4) Enter a description of your flowgram (perhaps "Convolution Exercise").

5) Enter <1> as the sampling frequency. We'll cover the meaning of the sample frequencies in a later lab.

6) Enter <3> for the number of display windows. (We have only marked three blocks for display, so we only need three windows).

7) Press <Y><Enter> to activate the auto display mode. The entire flowgram should be back on the screen.

8) Now the flowgram is ready to be run. Press <R>. While the flowgram is running, the block that is being executed will appear in reverse video.

9) After the run is completed, three graphs will appear on the screen. Are these plots what you expect?

Understanding and Using DSPlay's Graphs

Interpreting a DSPlay Plot:

The width of the displays is the block size we selected for the convolver, or 32 samples, and is the same for all plots (see appendix). For ease of viewing, the DSPlay program always scales a plot vertically so that it covers the entire area allotted to it. To accomplish this, the program assigns different scales to different graphs. Thus, one must be careful when comparing two plots.

For example, the triangle in the upper right plot is of height 5 while the pulses are only one unit high. This is not at all evident from a casual look at the graph because the axes are not even labeled!

Fortunately, there are other methods for viewing data with this software.

Using the Display Command:

1) Press the space bar to exit the window display and to return to the flowgram screen.

2) Use the <Up Arrow> to exit the "Home" mode and return to the flowgram. Position the cursor over the Convolver block.

3) Press <D> for the display menu and then <B> to pick the "plot Buffer" option. The "buffer" of a block contains the most recent output generated by a "run". (A shortcut is to hold down the <Alt> key

and press <D> at the same time). Two graphs should now fill the screen and your screen should look like figure 2. These are the plots of the real and imaginary components of the complex output of the convolver.

Note: Since both of the input signals to the convolver were real, the output of the convolver should also be purely real. The output is complex, however, because the program does the convolution in the frequency domain and then converts back into the time domain[2]. The imaginary part of the output is a result of slight computer error building up during the computations. Look carefully at the scales of the two graphs. The upper graph (real data) peaks at 5 which is just what we expected. The imaginary data (the lower graph) has a peak of about 1(f). The (f) stands for femtos and is very, very small (0.000000000000001). Thus, although it is not obvious at all from

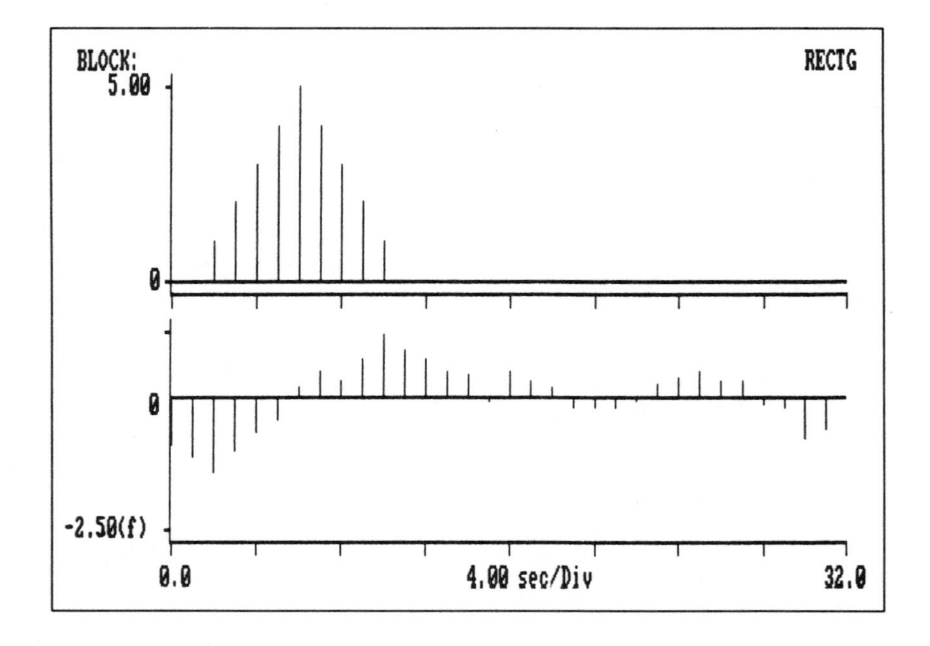

Figure 2. The output of the convolver of figure 1.

[2] This will be covered in a later lab.

the graph, the imaginary part of the output of the convolver is zero.

4) Try experimenting with the various commands on the bottom and top command lines.

s: This key changes the style of the plot.

v: Pressing the <v> key toggles the background of the plot. The light background is useful for making printouts of the screen.

b: This key puts you in "Buffer Mode". Buffer mode gives the index, time, and value of individual points (these are the numbers above the plot.) Using <right arrow> and <left arrow> you can move around the plot, each time getting at the top a display of the exact numerical value at the cross-hairs in the plot. To return to plot mode, press <p> (note that the top left of the screen reminds you of how to get back to plot mode).

z: Magnifies a section of the plot. The arrow keys can be used to adjust the position and size of the section to be magnified. Typing <Enter> will generate the magnified plot. Typing <r> will get you back to the original plot.

r: Replots the screen.

5) Press <Esc> to return to the flowgram and then use the Display command to look at the other two blocks.

Adding, Removing, and Changing Blocks

Modifying a block

1) Verify that you are in block mode, and move the cursor to one of the pulse generators.
2) Press <Enter> to get the Pulse parameter menu.
3) Change the pulse width to a new value.
4) Keep pressing <Enter> until you are back in flowgram mode.
5) Run the flowgram (Press <Home> and then <R>)
6) Use the Display command to look at the results of the convolution.
7) Experiment with various pulse sizes using the above procedure.

Changing a Block

The following procedure changes a Pulse block into a Waveform block.

1) Return to block mode and position the cursor over one of the Pulse blocks.
2) Press the <Del> key (found on the numeric key pad on some machines) and then the <y> key. This will delete the Pulse parameters and leave the block empty. If the <Del> key is pressed a second time, the empty block will be completely deleted.
3) Press <Enter> to get the Function Menu, and then press <I> to get the Input menu.
4) Choose the Waveform input function. The Waveform block generates its signal by linearly connecting the points given as parameters.
5) Enter a block name and enter <N> for cyclic (we only want a single impulse).
6) For the first point, enter an amplitude of 1. For the second point, enter 1.0 for the time (this is in seconds) and 0.0 for the amplitude. Press <Enter> for the rest of the points (3 through 12).
7) Now run your flowgram with the new impulse block.

Deleting Parts of the Flowgram

To delete a block, move the cursor (the flashing box) over the block and press the <Del> key twice.

To delete a line, move the cursor (again, the flashing box) over the block that is the source of the line. (The line connects to the left side of this block.) Press the <Ins> key to get into line mode (the cursor should be small, dark, and over a line). Position the small dark cursor over the line to be removed and press the <Del> key. This will remove the line.

To clean up and redraw the screen, press the F8 key. This is sometimes necessary after a delete.

To delete the entire flowgram, press the home key to get into flowgram mode (type <Home>) and press the <Del> key.

Problems:

Problem 1: *Convolution of square pulses*

Convolve a square pulse of length 6 with a square pulse of length 3.

a) Make a "picket fence" printout of the result (as in figure 3). If you do not have a printer available, make a simple sketch. Label the values of each of the points.

b) What is the length of the resulting signal?

c) How is this length related to the lengths of the two square pulses that were input into the convolver?

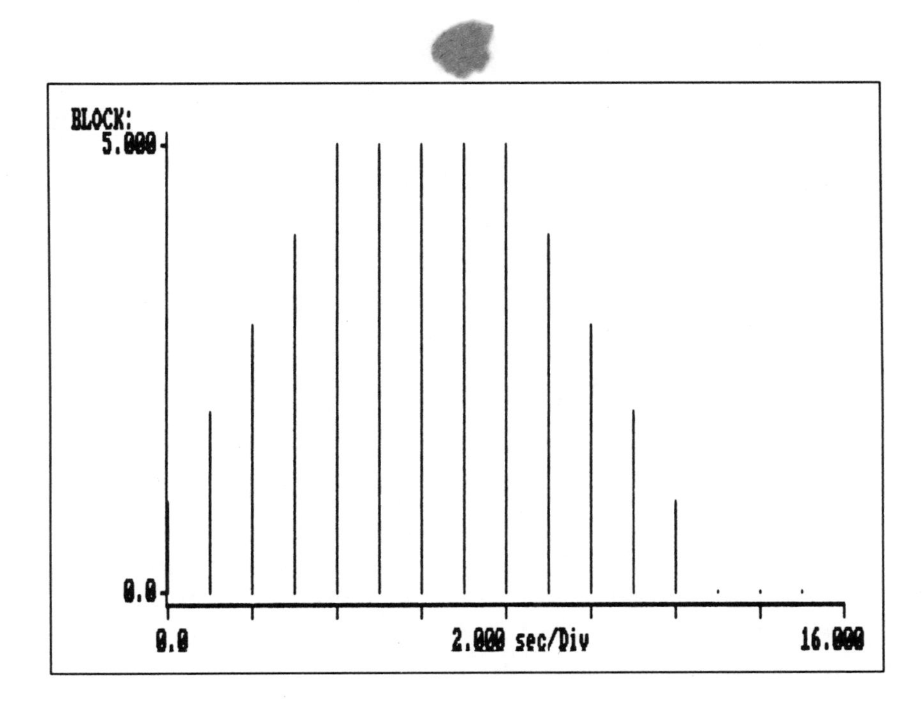

Figure 3. An example of a "picket fence" printout.

Problem 2: *An Exponential Sequence*

The exponential sequence is not part of the DSPlay Input Function Menu. Thus, we will have to build an exponential sequence using the blocks that are available.

To do this, we can use the following relationship:

$$a^n = e^{\ln a^n} = e^{n(\ln a)} \tag{1}$$

The exponential and natural log functions are both supported by the software.

This approach is somewhat limited because a must be positive (the log function is only defined for positive arguments). In the second lab, we will generate the sequence a^n for positive and negative a (and incidentally, it will be easier than the technique in this lab).

The sequence $x_n = n$ is called a ramp function. A ramp is simply an integrated step function. It can be generated by running a DC signal into an integrator. The "Integral" function is inside the "Non-lin" menu (in block mode, point to an empty block, type <Enter>, and select "Non-lin"). Build and verify a ramp generator before continuing.

The constant $\ln a$ can be viewed as a DC signal, and can be generated by running a DC signal of value a into an Ln block.

Build the rest of the exponential sequence. (The "Exp" and "Ln" function menus are both found in the "Non-lin" menu. The "Multiply" block is in the "Arith" menu.)

a) Make a printout (or simple sketch) of the flowgram of your exponential sequence. Mark on your printout or sketch the block that contains the value of a.

b) Perform the convolution of a square pulse of width 10 with an exponential sequence with $a = 0.9$. Since the exponential is a long sequence, it requires a long buffer length for the convolver. Use a buffer length of 128 for your convolve block. Plot or sketch your result.

Laboratory Exercise 2

The Z Transform and System Functions

Purpose:

DSPlay is used to find the inverse Z transforms of rational polynomials in z. The IIR filter block in DSPlay is used even though IIR filter structures will not have been explained in the course at this time. It is used as a general implementation of a rational system function. A much simpler solution to the problem of generating an exponentially decaying sequence is now possible by computing the inverse Z transform of a first-order, one-pole Z transform.

Background Material:

This lab assumes the student is familiar with the Z transform, the inverse Z transform, and the concept of a system function. In addition, there is a simple problem introducing difference equations. These topics are covered in:

Oppenheim and Schafer [1988]: Section 2.5, chapter 4

Oppenheim and Schafer [1975]: Sections 2.0 - 2.4

Jackson: Chapter 3 and Sections 4.1 - 4.2

Proakis and Manolakis: Chapter 3

Strum and Kirk: Chapter 6

Roberts and Mullis: Chapter 3

DeFatta, Lucas, and Hodgkiss: Sections 3.0 - 3.4

Kuc: Section 2.8, chapter 5

Rabiner and Gold: Sections 2.13 - 2.17

Oppenheim and Willsky with Young: Chapter 10

Example Procedures:

Using the IIR Block to Enter Z Transforms

While IIR filters will be covered in much greater detail later on, here we will use them to invert the Z transforms of signals. The IIR Block has one input and one output and contains a system function. This function is specified by entering the coefficients of its Z transform. If the Z transform of the system function is $H(z)$ then the IIR block computes the following:

$$output = input * h(n) \tag{2}$$

where "output" is the output of the IIR block, "input" is the input to the IIR block, and $h(n)$ is the inverse Z transform of $H(z)$. If the input to the IIR block is an impulse, the output of the IIR block will be the inverse Z transform of the system function. This is an excellent way to generate certain signals in DSPlay.

The parameters of the IIR block are the coefficients A_i and B_i, where

$$H(z) = \frac{A_0 + A_1 z^{-1} + A_2 z^{-2} + \cdots + A_{20} z^{-20}}{1 + B_1 z^{-1} + B_2 z^{-2} + \cdots + B_{20} z^{-20}} . \tag{3}$$

If a Z transform of order smaller than 20 is desired, then the last few coefficients are zero. If a Z transform of order larger than 20 is desired, then it must be factored or expanded by partial fractions into parallel or cascade forms. Thus, to enter your system function, first write it in the form of (3) and enter in the coefficients.

Creating an Exponential Sequence

The assignment in lab 1 to create the sequence a^n required six blocks and only worked for positive values of a. However, any signal with a rational Z transform of finite order, which includes a^n for any a, can be generated using the IIR filter block.

1) Find the Z transform of the sequence $a^n u(n)$. Make sure that your resulting equation is in the form of (3).
2) With your favorite, stable value for a, use an IIR block and an impulse (Waveform or Pulse) block in DSPlay to create an exponential sequence. Check to make sure that the sequence is indeed $a^n u(n)$.
3) Try a negative value for a. Negative values were not possible using the methods of the previous lab.
4) Try a value for a the leads to an unstable signal.

Problems:

Problem 1: *Convolving Exponentials*

a) Let $h(n) = a^n u(n)$ and $x(n) = b^n u(n)$. Use a convolver block and the technique above to find and print (or sketch) $y(n) = h(n) * x(n)$. Let $a = 0.9$ and $b = ½$.

b) Now find $y(n)$ *without* a convolver block. Use a total of three blocks or fewer. Make a printout (or sketch) of your flowgram. On your printout or sketch, write the values of the non-zero coefficients of each IIR block used.

Problem 2: *Inverting a Z Transform*

Given the Z transform below:

$$H(z) = \frac{1 - 0.995z^{-1}}{1 - 1.99z^{-1} + z^{-2}} \tag{4}$$

find and make a printout (or simple sketch) of $h(n)$ using DSPlay.

Problem 3: *Creating a complicated sequence*

Find the peak values of the following sequences:

a) $h(n) = (0.95)^n \sin(0.1n)u(n)$.

b) $h(n) = (0.95)^n \sin(0.2n)u(n)$.

Hint: The Buffer mode or Tabulate buffer subcommand of the display command is very useful for finding the values of individual points in a graph.

Problem 4: *Implementing Difference Equations*

a) Let $y(n) = 2x(n) + 0.75y(n-1) + 0.125y(n-2)$. Find $H(z)$ so that $Y(z) = X(z)H(z)$.

b) Make a printout (or simple sketch) of $h(n)$

c) Find the first five values of $y(n)$ $\{ y(0), y(1), \cdots, y(4) \}$ when $x(n)$ is a square pulse of length five.

Assume $y(n) = 0$ for $n < 0$.

Laboratory Exercise 3

Pole/Zero Plots and Frequency Response

Purpose:

This lab explores the relationship between rational Z transforms, their pole/zero plots, and their frequency responses. The FFT block is used to estimate the frequency response of systems, even though the student is not expected to understand the mechanics of the FFT yet.

Background Material:

It is assumed that the student knows how to graphically estimate a frequency response from a pole-zero plot. This is subject is covered in:

Oppenheim and Schafer [1988]: Section 5.3

Oppenheim and Schafer [1975]: Section 2.4

Jackson: Sections 4.1 - 4.3

Proakis and Manolakis: Sections 3.3, 5.2, and 5.5

Strum and Kirk: Chapter 5
Roberts and Mullis: Section 2.5
DeFatta, Lucas, and Hodgkiss: Section 3.4.6
Kuc: Sections 5.10 - 5.11
Rabiner and Gold: Section 2.18

Example Procedure:

Using DSPlay to Determine the Frequency Response

1) Delete any flowgram currently in memory.
2) Build an Impulse and connect it to an IIR block with system function equal to $H(z)$ below:

$$H(z) = \frac{1 + z}{z - 0.80} \qquad (5)$$

(Remember to convert the function into the correct form before entering coefficients.)
3) Now build an FFT block to the right of the previous two blocks. The FFT block computes the Discrete Fourier Transform (DFT), a subject that will be covered later in the course. We are using the FFT here to compute the frequency response, and for the purposes of this exercise, we can assume it computes the discrete-time Fourier transform (DTFT), or frequency response. The FFT block is in the Spectrum section. Choose a large value for the length of the FFT buffer: 256 or 512 will do. A longer buffer gives a better approximation of the frequency response. (It also controls the number of samples generated by DSPlay; see the appendix). Connect this FFT block to the output of the IIR block.
4) Run this flowgram and display the output of the FFT block. The FFT block produces complex output. Thus, the display for the FFT block is two graphs: the upper is the real data and the lower is the imaginary data. The horizontal axis is frequency and runs from 0 to $2\pi/T$.

   Press the <C> key to toggle to a phase-magnitude plot. To find the specific values of points on the graph, try out the buffer mode while in display mode.

Problems:

Problem 1: *A Simple Low Pass Filter*

a) Sketch a pole/zero plot of the function in (5).

b) Make a *sketch* of the frequency response (magnitude and phase) of $H(z)$ using your pole/zero plot and the rules for graphically interpreting pole-zero plots. Check this sketch against the plots generated by DSPlay.

c) Try varying the position of the pole. Move it closer and/or further away from the unit circle. (Note: if you move the pole on, or beyond, the unit circle; you will produce an unstable system.) How does this affect the impulse response? How does moving the pole affect the magnitude response?

Problem 2: *The Unity-Gain Resonator*

A unity-gain resonator is a system with a peak gain of approximately unity at a frequency called the resonant frequency and attenuation at all other frequencies. An appropriate second order form is

$$H(z) = \frac{b_0(1 - z^{-2})}{1 + a_1 z^{-1} + a_2 z^{-2}} . \tag{6}$$

The peak gain of this system occurs roughly at ω_0, where ω_0 is the argument (angle) of the poles, assuming the poles are close to the unit circle. A pole/zero plot of a resonator is shown in figure 4.

a) Use graphical techniques to estimate and sketch the magnitude response of the system shown in figure 4.

b) Set $b_0 = 1$, and implement the system of part (a) using DSPlay. What are the values of a_1 and a_2?

c) What value of b_0 gives the system a peak gain of unity?

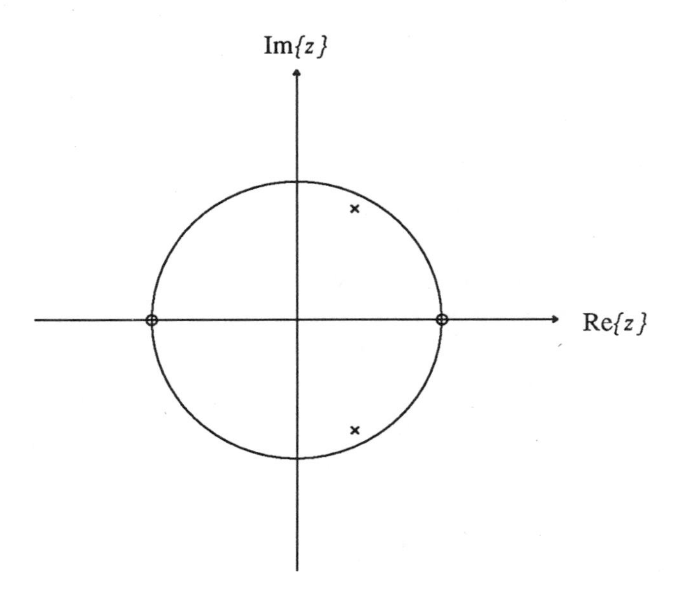

Figure 4. The pole/zero plot of the resonator from problem 2. The poles are located at $z = 0.4 + j0.8$ and $z = 0.4 - j0.8$.

Problem 3: *A Modified Resonator*

a) Use graphical techniques to estimate and sketch the magnitude response of the system shown in figure 5.

b) Implement this system in DSPlay by cascading two resonator systems. Printout (or sketch) your flowgram and indicate on your printout (or sketch) the values of the non-zero coefficients of each of the IIR blocks used.

Problem 4: *An Interesting System Function*

Using the system function:

$$H(z) = \frac{1 - az^{-1}}{1 - a^{-1}z^{-1}} \qquad |a| > 1 \tag{7}$$

a) Draw the pole/zero plot of $H(z)$ and make a quick sketch of the magnitude response. Try several values for a.

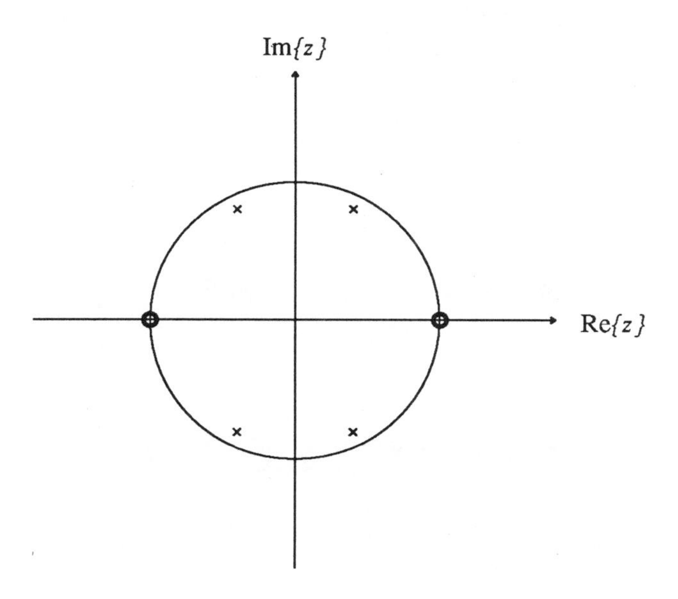

Figure 5. A modified resonator that peaks at two different frequencies. The poles are located at $z = -0.4 + j0.8$, $z = -0.4 - j0.8$, $z = 0.4 + j0.8$, and $z = 0.4 - j0.8$. There are double zeros at $z = +1$ and $z = -1$.

b) Use DSPlay to verify your sketch.

c) Show analytically that $H(z)$ has constant magnitude response. Find that magnitude.

Problem 5: *Another Interesting System Function*

Consider the system

$$H(z) = 1 - z^{-8}. \tag{8}$$

a) Sketch the (pole)/zero plot of $H(z)$, and use graphical techniques to estimate the magnitude response.

b) Verify your result using DSPlay. This filter is an example of a comb filter.

Laboratory Exercise 4

Linear Phase Filtering

Purpose:

This introduces linear-phase FIR filtering and demonstrates the difference between even-symmetric and odd-symmetric filters. The lab also looks at group delay and group symmetry. In addition to FIR filters, the lab introduces non-causal filtering using signal reversal to accomplish linear-phase filtering.

Background Material:

A brief introduction to FIR filtering and the concept of linear-phase are prerequisite. The material is covered in:

Oppenheim and Schafer [1988]: Section 5.7

Oppenheim and Schafer [1975]: Section 4.5.3

Jackson: Section 5.3

Proakis and Manolakis: Section 5.3.4

Strum and Kirk: Section 9.2

Roberts and Mullis: Section 6.2

DeFatta, Lucas, and Hodgkiss: Section 5.2

Kuc: Sections 3.4, 6.7

Rabiner and Gold: Sections 3.3 - 3.7

Oppenheim and Willsky with Young: Section 6.4 & Problems 6.7 - 6.9

Example Procedure:

Frequency Response of an Even Symmetric FIR Filter

In DSPlay, FIR filters are implemented with the FIR block found in the Filter menu. Although not labeled, the 128 filter coefficients run from left to right, top to bottom and are in the following form:

$$H(z) = A_0 + A_1 z^{-1} + A_2 z^{-2} + A_3 z^{-3} + \cdots + A_{128} z^{-128} . \qquad (9)$$

Thus, A_0 is the first coefficient and A_1 is the second, etc. The system is shown in figure 6. When done entering coefficients, press the <Ctrl> key and the <Enter> key at the same time. This tells DSPlay that you are done with the block and is much faster than holding the <Enter> key down con-

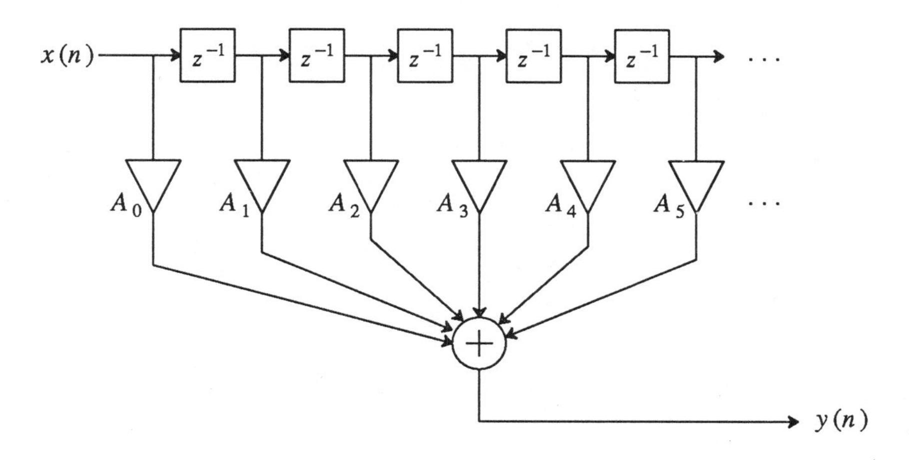

Figure 6. Implementation of an FIR Filter in DSPlay

tinuously to accept all the unused parameters.

The FIR block works just as the IIR block does: $output = h(n) * input$. In fact, the IIR block may be used to enter short FIR filters (simply enter zero for all of the denominator coefficients.)

a) Build an FIR filter with even symmetry. Any FIR filter will do as long as $A_m = A_{M-m}$ where A_M is the last tap. See figure 7 for an example of an even-symmetric FIR filter. (Figure 5.11 of Jackson has two more examples of even-symmetric FIR filters.) Note that an even-symmetric FIR filter may have an even or odd number of taps.

b) Run an impulse into your FIR filter, and run the output of the FIR into an FFT with a long buffer length.

c) Look at the phase response of the filter. If all has gone well, this should be linear. The magnitude and phase response of the even FIR filter of figure 7 is shown in figure 8. Note that the phase response in figure 8 is indeed linear.

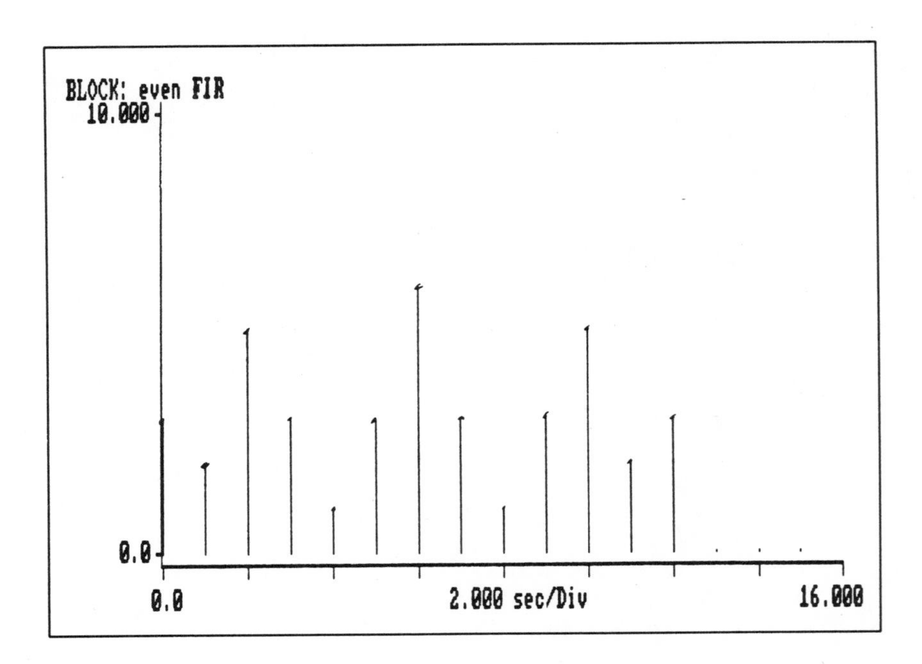

Figure 7. An even-symmetric filter.

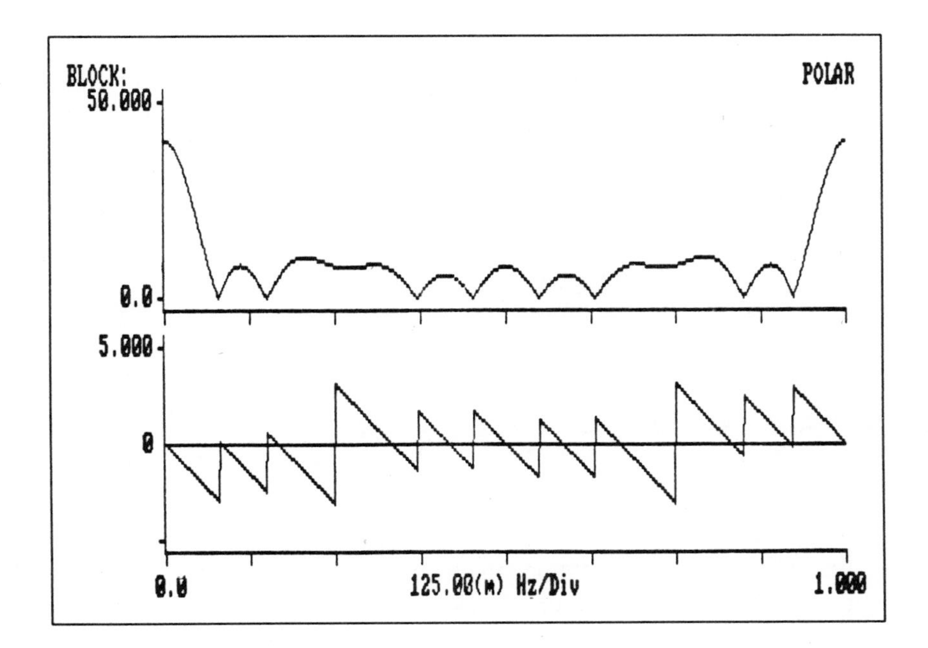

Figure 8. Magnitude and phase response of the filter in figure 7.

d) Now enter new FIR coefficients, always keeping even symmetry. Try long and short FIR filters. Also try FIR filters with odd and even numbers of taps (coefficients).

Problem 1: *Group Delay and Group Symmetry*

a) Using your favorite even-symmetric FIR filter from the example procedure above, filter a square pulse. Sketch or print your input pulse and the output of your filter.

b) What is the group delay (in seconds) of the filter? The group delay is the time between the centers of symmetry of the input and the output signals.

c) How is the group delay related to the length of your filter?

d) How is group delay related to the slope of the phase response of the filter?

Problem 2: *Odd Symmetric FIR Filters*

a) Build an FIR block with odd-symmetric taps. Recall that odd symmetry means that $A_m = -A_{M-m}$.

b) Find the phase response of this filter and compare it to the phase response of an even-symmetric FIR filter. What is the difference between the two?

c) Filter a cosine wave with your odd-symmetric FIR filter. A good looking cosine wave has a frequency around 0.1 Hertz. What is the output?

Problem 3: *Linear Phase Through Signal Reversal*

Although linear phase is easy to achieve with an FIR filter, it can also be achieved with *any* filter using signal reversal. The input signal is run forward and then backward through the same filter. Obviously, this operation is not causal, and can only be easily applied if the input signal is finite in length.

Given an input $x(n)$ and a filter $h(n)$, the technique is as follows:

$$\begin{aligned} g(n) &= h(n) * x(n) \\ r(n) &= h(n) * g(-n) \\ y(n) &= r(-n) \end{aligned} \tag{10}$$

where the output is $y(n)$.

a) Let $f(n)$ be such that $y(n) = f(n) * x(n)$. What is $f(n)$ in terms of $h(n)$?

b) If $h(n)$ is causal, will $f(n)$ also be causal?

c) Find the frequency response $F'(\omega)$ and express it in terms of $|H'(\omega)|$ and $\angle H'(\omega)$. It will help if you assume that $h(n)$ is real.

Problem 4: *Implementing Signal Reversal in DSPlay*

All signals start at time $t = 0$ in DSPlay. Therefore, it is impossible to generate signals such as $g(-n)$ because the program cannot handle the negative index. To avoid the negative index, we can displace the signal. Using an appropriate delay, the input may be forced to start at a new reference point. By delaying the input, it is possible to handle non-causal systems.

Signal reversal ($g(n) \rightarrow g(-n)$) is implemented by the "reVerse" block in the "data" category.

a) In DSPlay, build a linear-phase filter system using the signal reversal technique described above. For your $h(n)$, use an IIR filter that has an exponential sequence as its impulse response. The default frame length is 128[3], so to get your reference point at the center of this frame, use a delay of 63 (the delay block is in the "control" category). Print or sketch the flowgram for your linear-phase filter using signal reversal.

b) Use an impulse as your input to find the impulse response of the system. Is the impulse response symmetric?

c) Since the input is delayed, the output will also be delayed, causing a large linear phase component to be added to the phase of a non-delayed $y(n)$. As a consequence, the phase will be "wrapped around". Taking this into account, does the phase response match your theoretical phase response calculated in problem 3?

   The effect may be difficult to see because the output of the FFT is a sampled frequency response and the resolution may be inadequate. To get more resolution, use the "zero pad" block (in the "data" category) to convert from a 128-point frame size to a 512-point frame size. Then attach a 512 point FFT to the zero pad block. The 512 point FFT will "fill in" the frequencies. The theory behind this will be explored in a future lab.

---

[3] The output frame length or buffer length is discussed in the appendix.

Laboratory Exercise 5

Amplitude Modulation

Purpose:

As an application of filter design, an interesting signal is generated, modulated, filtered to remove one of the sidebands, demodulated, and low-pass filtered. If the demodulation frequency is different from the modulation frequency, then a spectral shift results. This effect is demonstrated.

Background Material:

The relationships between the time domain and the frequency are covered in:

Oppenheim and Schafer [1988]: Chapter 2

Oppenheim and Schafer [1975]: Chapter 2

Jackson: Chapter 3

Proakis and Manolakis: Section 6.1

Strum and Kirk: Chapter 4

Roberts and Mullis: Sections 4.1 - 4.2

DeFatta, Lucas, and Hodgkiss: Through section 3.4

Kuc: Chapters 2 - 3

Rabiner and Gold: Sections 2.1 - 2.9, and 2.16

Oppenheim and Willsky with Young: Chapter 7 (modulation)

Problem 1: *Basic Cosine Modulation*

Build the Amplitude Modulation section of figure 9, generating the signal $y(n)$.

Create the modulating signal, $x(n)$, by running an impulse into an FIR filter with the following coefficients: -2.0, -3.9, -5.3, -4.4, -1.0, 3.9, 8.3, 10, 8.3, 3.9, -1.0, -4.4, -5.3, -3.9, -2.0, 0, 0, ... This filter is stored on the disk as: xlab5. To load it into a flowgram, do the following: 1) Create and empty block; 2) Move the "flashing box" cursor over the empty block; 3) Press LP for "load parameters"; 4) Enter xlab5 as the file name.

Look at the FFT of the resulting signal (a block size of 256 is recommended).

a) What is the bandwidth of $x(n)$? (ie. What is the width, in Hertz, of one of the frequency "lumps" of $x(n)$?)

b) Now multiply (modulate) $x(n)$ with a 0.25Hz cosine carrier wave. Look at the FFT of the result, $y(n)$. What is the bandwidth of this signal?

c) What is the largest possible bandwidth for $x(n)$ so that $y(n)$ will not have any aliasing distortion?

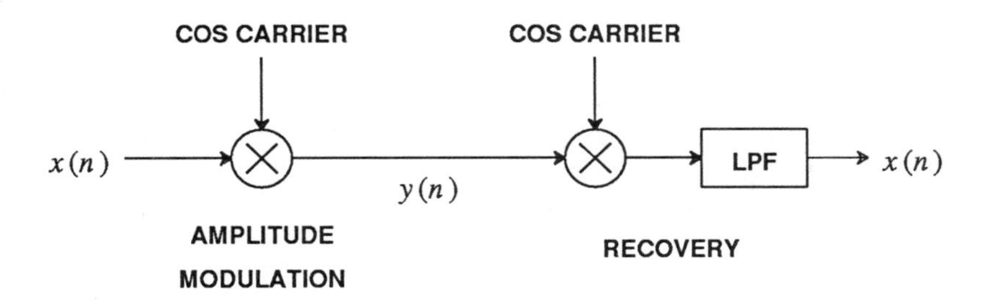

Figure 9. Double-sideband modulation and demodulation system.

d) Change the cosine carrier wave frequency to 0.15Hz. Is there any distortion visible in $|Y'(\omega)|$ due to overlapping sidebands?

e) With the carrier wave frequency set to 0.15Hz, what is the largest possible bandwidth for $x(n)$ so that $y(n)$ will not have any distortion?

Problem 2: *Recovery of Modulated Signal*

The modulated signal created in problem 1 may be recovered by multiplying it again with the same cosine carrier wave and then passing it through the low-pass filter with a cutoff at 0.25Hz, as shown in figure 9.

A simple FIR low-pass filter with a cutoff of 0.25Hz has the following coefficients: 0.076923, 0, -0.09091, 0, 0.111111, 0, -0.14286, 0, 0.2, 0, -0.33333, 0, 1, 1.570796, 1, 0, -0.33333, 0, 0.2, 0, -0.14286, 0, 0.111111, 0, -0.09091, 0, 0.076923. This filter is also stored on disk. Its file is: lplab5.

Try this using DSPlay and a cosine carrier wave of 0.25Hz. Look at the FFT of each stage of your results.

a) What is the range of cut-off frequencies that could have been used for the low-pass filter?

b) Change the carrier frequency to 0.21Hz. Make sure that both the modulator and demodulator frequencies are the same. Does the system still work for this new carrier frequency?

c) Change the cosine carrier frequencies to 0.15 Hz. Is the output of the system still $x(n)$?

Problem 3: *Single-Sideband Modulation*

In problem 1 it was noted that the modulated signal used twice as much bandwidth as the original signal, $x(n)$. Clearly, there most be some redundancy in the modulated signal.

A method for removing this redundancy is called single-sideband modulation. Both upper sidebands (the frequency 'lumps' closest to $\omega = \pi$) or both lower sidebands (the frequency 'lumps' close to $\omega = 0$ *or* 2π) may be removed without any loss of information.

A system which implements this is illustrated figure 10.

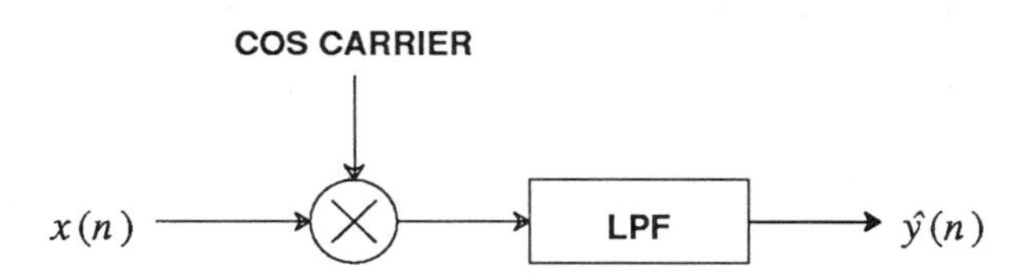

Figure 10. A single-sideband modulation system.

Implement this system using DSPlay and a cosine carrier wave of 0.25Hz.

a) What is the bandwidth of $\hat{y}(n)$?

b) In order to take only one set of sidebands, what should be the relationship between the the frequency of the carrier and the cutoff frequency of the low-pass filter?

c) Try using the same recovery scheme that is used in figure 9. Is the output still $x(n)$?

Problem 4: *Varying Carrier Frequencies*

In the real world, different generators are often used for the modulation carrier and the recovery carrier. If the two carriers are significantly different, errors will occur.

Build a single-sideband system in DSPlay using two different cosine generators for the modulating and recovery carriers. Set the modulating carrier to 0.25Hz but set the recovery carrier to 0.19Hz.

a) What happens to the frequency spectrum of the output signal?

b) Now set the recovery the recovery carrier to 0.30Hz. How does this affect the frequency spectrum of the output signal?

A system like this one can be used to shift the frequency spectrum of a signal. The drawback of this system is that it requires a high sampling rate. A system which produces a spectral shift without the high sampling rate is discussed in the next lab.

Laboratory Exercise 6

Spectral Shifter

Purpose:

The purpose of this lab is to gain familiarity with complex-valued signals and systems, analytic signals, and phase splitters. Complex-valued time-domain signals are used. If these are not being covered in the course, this lab should be skipped. The objective is to construct a spectral shifter (like that in lab 5) using complex-valued signals. This method admits a much lower sampling rate than the method used in lab 5.

Background Material:

This lab requires a familiarity with the basic operations involving complex numbers. This may be found in a math text or in the following texts:

Oppenheim and Schafer [1988]: Chapter 10

Oppenheim and Schafer [1975]: Section 7.4

Strum and Kirk: Section 7.2 and Appendix A

DeFatta, Lucas, and Hodgkiss: Sections 4.9.2 and 2.1.3

Kuc: Section 2.2

Oppenheim and Willsky with Young: Sections 2.4.2 - 2.4.3 and 7.1

Problems:

Given a signal $x(n)$ with spectrum shown in figure 11, the objective of this lab is to shift the spectrum of the signal by an amount a, as shown in figure 12.

Problem 1: *A First Attempt Spectral Shifter*

The input signal $x(n)$ has the following time domain representation:

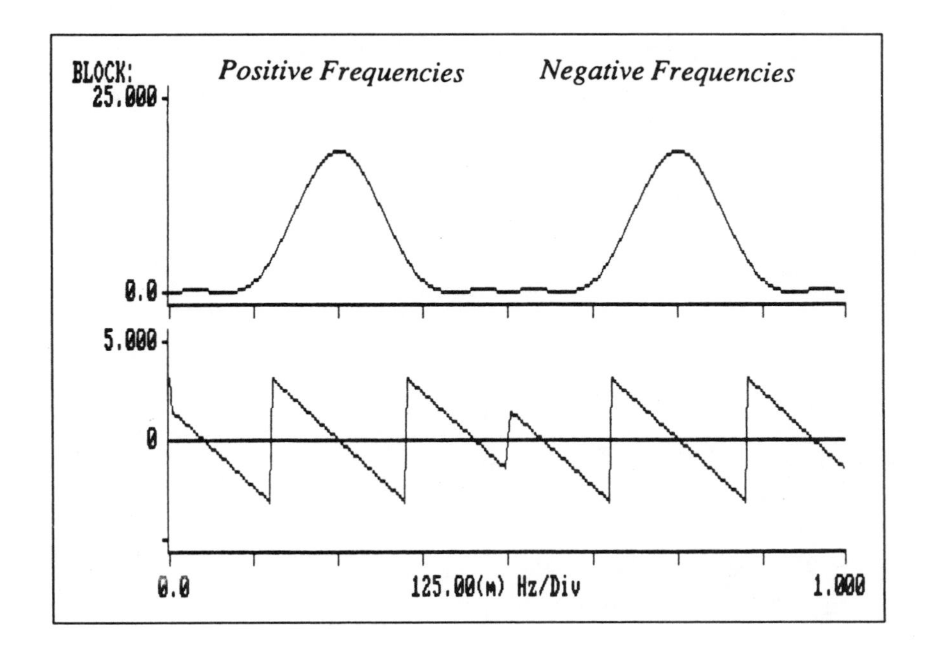

Figure 11. $X'(\omega)$: The magnitude and phase response of $x(n)$ given in (11).

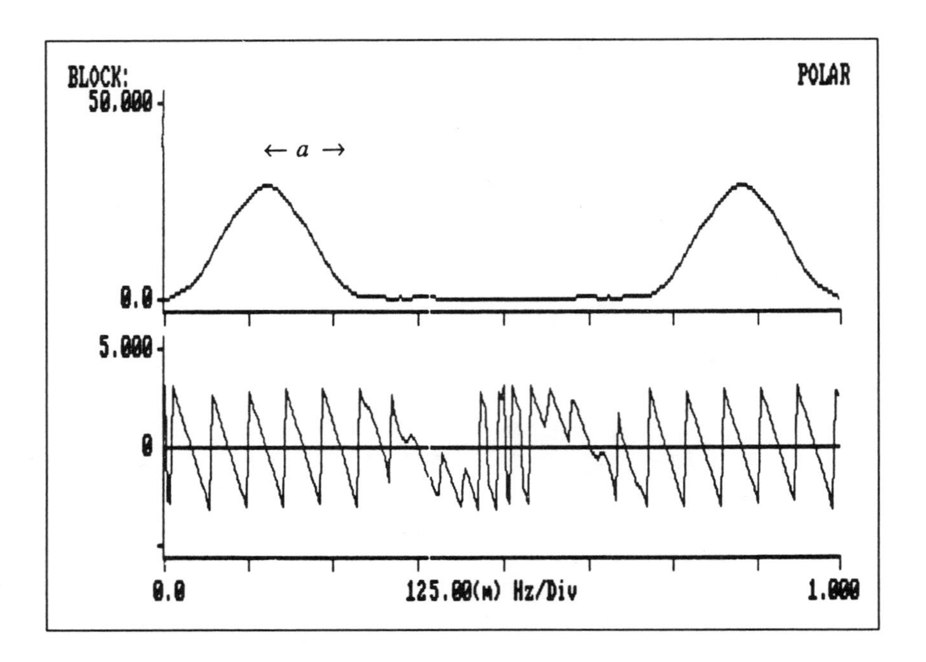

Figure 12. $Y'(\omega)$: The magnitude and phase response of $y(n)$, a spectrally shifted version of $x(n)$.

$$x(0)=1, x(1)=0, x(2)=-3, x(3)=0, x(4)=5, x(5)=0,$$
$$x(6)=-5, x(7)=0, x(8)=3, x(9)=0, x(10)=-1, x(11)=0\,. \quad (11)$$

This signal can be created using the waveform block. It is also stored on the disk as: xlab6. To load it into a flowgram, move the "flashing box" cursor over an empty block and press LP for "load parameters" and use xlab6 as the file name. Build $x(n)$ in DSPlay. It should have the spectrum shown in figure 11.

A first attempt might be to modulate the signal with a cosine. Print or sketch the magnitude of the frequency response of $x(n)\cos(naT)$ to show that simply multiplying by a cosine carrier will not give the desired results. Let $a = 0.05Hz$.

Problem 2: *Another Attempt at a Spectral Shifter*

Another simple approach is to modulate the signal with a complex exponential.

Fortunately, DSPlay can handle complex signals, although the treatment is somewhat different. As a result of this, one must be careful not to mix the two types of signals. For example, the DSPlay Add Block can add two real signals together or it can add two complex signals together, but it cannot add a complex signal to a real signal. This is true for other blocks in DSPlay as well (ie. Multiply, Convolve, Subtract, etc.)[4]

DSPlay provides two blocks to convert data from real to complex or from complex to real. Both of these blocks are found in the "Data" category.

a) Build a complex exponential sequence in DSPlay. Recall that $e^{j\omega nT} = \cos(\omega nT) + j\sin(\omega nT)$. Print or sketch the flowgram used to create your sequence. Use the FFT block to verify that the spectrum of this signal is as expected.

b) Use your new complex exponential sequence to modulate the signal x(n). Remember that the multiplier needs two complex inputs. Thus, x(n) will have to be converted into a complex signal. Print or sketch the frequency response of the modulated signal. Explain why this also does not produce the desired signal.

Problem 3: *Creating Analytic Signals Using a Phase Splitter*

The simple spectral shifter of problem 2 correctly shifts the positive frequency components of $x(n)$, but not the negative frequency components (which are plotted on the right half of the screen by DSPlay). The solution is to *remove* these complicating negative frequencies by filtering. This requires a filter that is not conjugate-symmetric about zero frequency, and hence cannot have a real-valued impulse response.

A signal with only positive frequencies is called an *analytic signal* and it is a complex signal in the time domain. We can get an analytic signal from a real-valued signal using a *phase splitter*. The impulse response of a phase splitter can be obtained by modulating the impulse response of a

[4] Unfortunately, DSPlay may not always warn the unsuspecting user of this shortcoming. Thus, it is very important to make sure that all of your blocks with two inputs are receiving two real signals or two complex signals.

low-pass filter with a complex exponential. Use the low pass, FIR filter from lab 5. This filter is stored on the disk as: lplab5. The filter may be loaded into a flowgram by moving the "flashing box" cursor over an empty block, pressing LP for "load parameters" and entering lplab5 as the file name. Modulate the impulse response of this low pass filter with your complex exponential.

The resulting signal is the impulse response of a phase splitter. The frequency response of this filter is shown in figure 13. Note that the frequency response is not symmetric about zero.

An easy way to now implement the phase splitter is to convolve this impulse response with a signal that we wish to filter. The output of the convolver will be an analytic signal.

a) In DSPlay, build a phase splitter and generate $\hat{x}(n)$, the analytic version of $x(n)$. **Hint:** Be sure to pick the frequency of the complex exponential correctly.

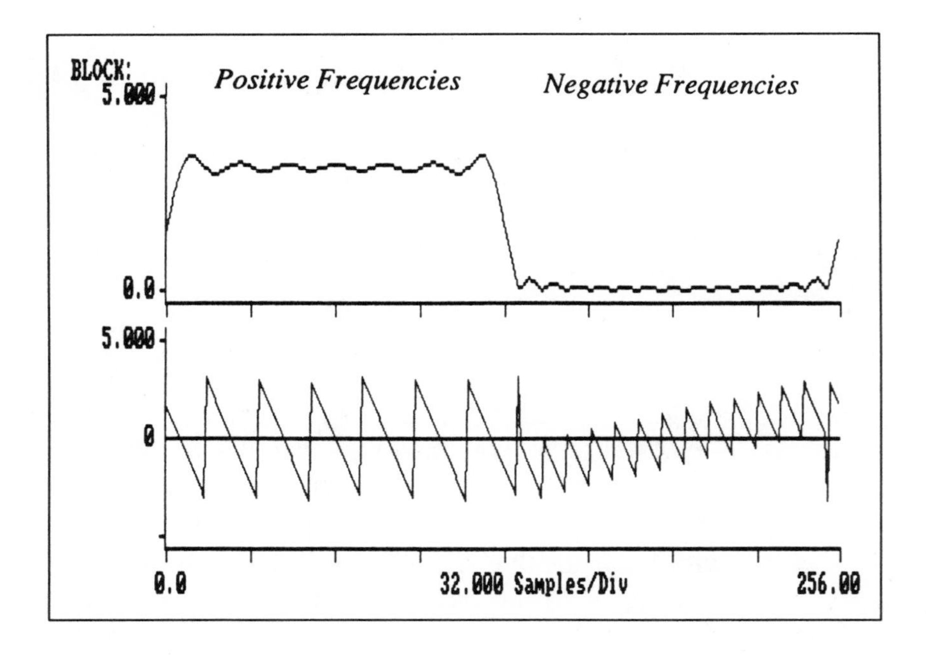

Figure 13. A phase splitter created from the 0.25Hz low-pass filter by modulating its impulse response with a complex exponential.

b) Print or sketch your flowgram.

Problem 4: *Building a Spectral Shifter*

Now we have all of the tools necessary to build a spectral shifter. The strategy is as follows:

First, use the phase splitter to eliminate the negative frequencies of the input signal, $x(n)$, and to create an analytic signal, $\hat{x}(n)$ (figure 14).

Second, modulate $\hat{x}(n)$ with a complex exponential to shift the new analytic signal in frequency. This shifted signal, $\hat{y}(n)$ will also be an analytic signal (see figure 15).

Third, take the real part of $\hat{y}(n)$. This new signal, $y(n)$, will be the desired shifted version of $x(n)$ (see figure 12 for the frequency response of $y(n)$).

a) Build a spectral shifter as discussed above. Use $a = -0.06\ Hz$.[5]

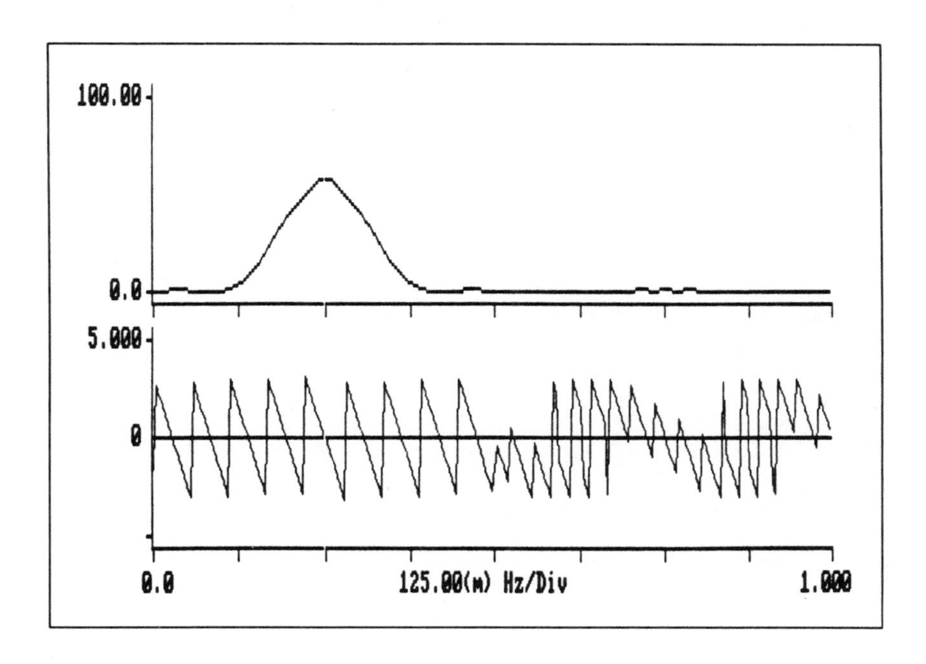

Figure 14. $\hat{X}'(\omega)$: Magnitude and phase response of $\hat{x}(n)$.

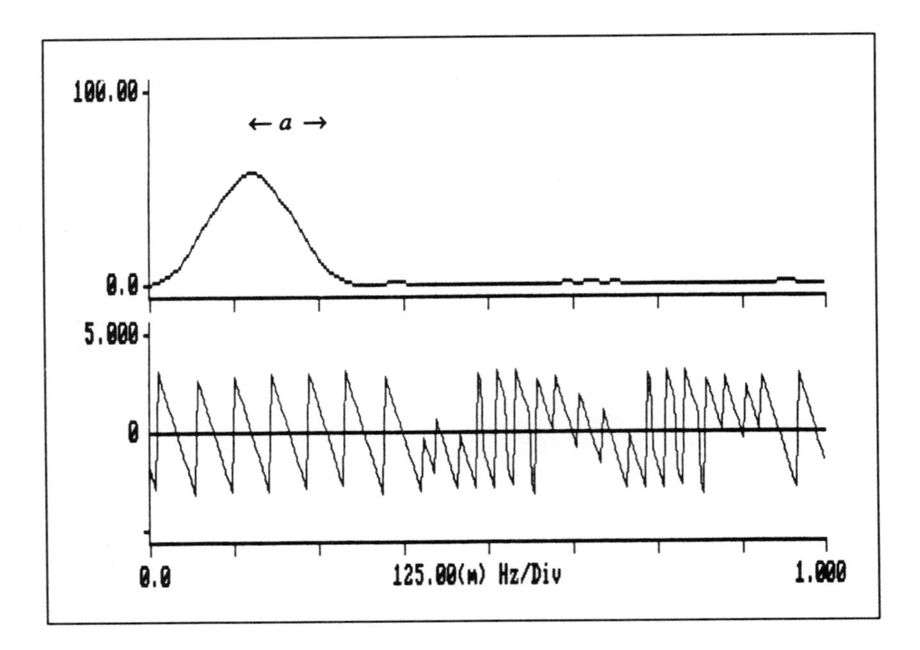

Figure 15. $\hat{Y}'(\omega)$: Magnitude and phase response of $\hat{y}(n)$.

b)
Print or sketch your spectral shifter flowgram. (You may indicate the phase splitter with a single block if you desire.)

c) Try a 0.1Hz sine wave as the input to your system. What is the frequency of the output signal?

---

[5] Note that DSPlay does not allow for negative frequencies for sine and cosine. To get around this problem, the following relationships may be helpful: $sin(-\omega) = -sin(\omega)$ and $cos(-\omega) = \cos(\omega)$. Also note that this value of a is different than the a in the figures.

Laboratory Exercise 7

The DFT, FFT and Circular Convolution

Purpose:

The DFT and circular convolution are used, explained, and explored. The use of zero-padding and the impact of DFT length are emphasized, since these ideas are crucial to proper use of the tools.

Background Material:

The DFT and FFT are covered in the following:

Oppenheim and Schafer [1988]: Chapters 8 - 9

Oppenheim and Schafer [1975]: Sections 3.5 - 3.8 & 6.2

Jackson: Sections 7.0 - 7.4

Proakis and Manolakis: Sections 4.5, 5.4, and 9.1 - 9.4

Strum and Kirk: Chapter 7

Roberts and Mullis: Sections 4.5, 4.8, 5.1, and 5.2

DeFatta, Lucas, and Hodgkiss: Chapter 6

Kuc: Chapter 4

Rabiner and Gold: Sections 2.21 - 2.24, 6.1 - 6.3, 6.11 - 6.12, 6.14

Procedure and Problems:

Problem 1: *The DFT and the FFT*

DSPlay provides both an FFT and DFT block. It is important to understand the difference. The FFT is simply a fast way to calculate the DFT. It uses the decimation in time algorithm, which requires about $N \log_2 N$ multiply-add computations. The DFT block, however, directly uses the summation that defines the DFT. This takes about N^2 computations.

The following table shows the speed difference between the two methods:

Size	*DFT Block*	*FFT Block*	*Speed–Up*	
N	N^2	$N \log_2 N$	*Factor*	
128	16,000	900	18×	(12)
256	65,500	2,000	32×	
512	262,000	4,600	57×	

Because of the large difference in computation time, the DFT block is rarely used. Its only advantage is that it can compute a subset of values in the frequency domain. This can save time when only one or two samples of the DFT are required.

To see for yourself the difference in computation time, connect a square pulse to both a 128 point FFT block and a 128 point DFT block. To make the comparison fair, we want the DFT block to also produce a full spectrum. Thus, use 0 for the low limit of the DFT block and 127 for the high limit.

Since the DFT Block is so slow, we will use the FFT Block from now on to calculate our DFTs.

a) Find the 8 point FFTs of each of the four signals in figure 16.

   Ignoring the slight computer round-off error, which of these FFTs are purely real?

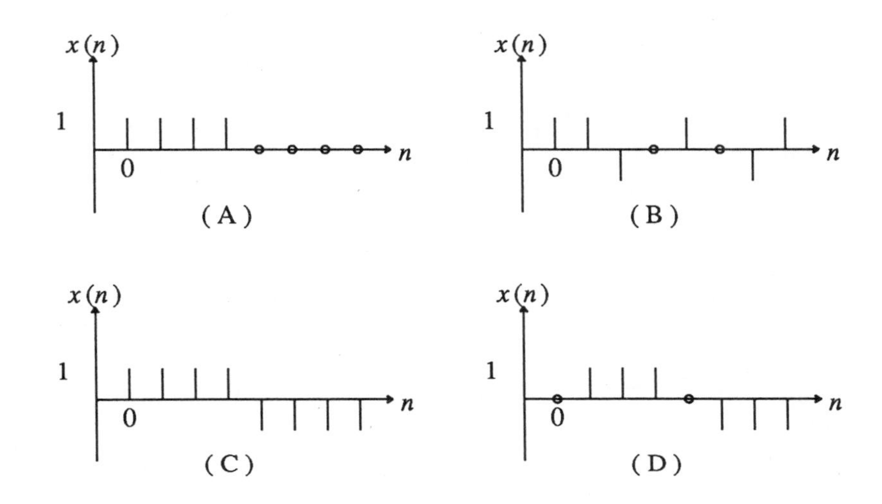

Figure 16. Four signals used in problem 1.

b) Which of these four FFTs are purely imaginary?

Problem 2: *The Effects of DFT Length*

Let $x(n)$ be defined as follows:

$$x(n) = \begin{cases} 1, & \text{if } n = 0, 1, 2, \textit{or } 3 \\ 0, & \textit{otherwise} \end{cases} \tag{13}$$

a) Find and print or sketch the 4 point DFT of $x(n)$.

b) Now look at the 8, 16, 32, and 64 point DFTs of $x(n)$. Note how the frequencies are "filled in" as the size of the DFT is increased. In other words, as we compute longer DFTs, we get more samples of the discrete-time Fourier transform of the original pulse.

c) Find the complex value of the discrete-time Fourier transform $X'(\omega)$ at $\omega = \frac{9}{128}\frac{2\pi}{T}$.

Problem 3: *Properties of the DFT*

Compare the 128 point FFT of a 0.125Hz cosine wave to the 128 point FFT of a 0.123Hz cosine wave. It will be easier to observe the differences if the FFTs are first converted to polar (Magnitude and Phase) plots.

Even though the inputs to both FFTs are cosine waves, the shapes of the FFTs are different. Why is this so?

Problem 4: *Zero Padding*

The Zero pad block in DSPlay adds trailing zeros to its input. Thus, the length of the output of the Zero pad block is longer than the length of its input. Here, we are interested in "filling in" the frequencies of the DFT of a cosine wave just as we "filled in" the frequencies of $x(n)$ in problem 2.

Connect a Zero pad block to the output of a 0.125Hz cosine wave. Make the input frame length = 128, so that we will be looking at the same cosine wave as in problem 3. In order to do a good job of "filling in" the frequencies, make the output buffer length = 512. Now connect the output of the Zero pad block to a 512 point FFT.

a) Look at the output of the 512 point FFT. Does it look the same as the 128 point FFT from problem 3?

b) Now change the cosine block to 0.123 Hz but keep the rest of the flowgram unchanged. Look at the output of the 512 point FFT. Is the shape of this FFT very different from the FFT of the "filled in" 0.125Hz cosine? Why or why not?

c) Now find the 512 point FFT of a non-cyclic pulse[6] with a width of 128. How is the output of this FFT related to the output of the FFT in part (a)?

d) Multiply the pulse block from part (c) with a new 0.125Hz cosine block[7]. Look at a 512 point FFT of the output of the Multiply block. How is this signal related to the output of the FFT of part (a)?

---

[6] Even though the pulse is non-cyclic, DSPlay insists on being given a period. A period of 2000 will work nicely

[7] Do not use the cosine block from part (a) as this will result in a "buffer size" error.

Problem 5: *Windowing*

The strange shape of the FFT in the problem above is due to the truncation of the cosine wave after 128 points. This truncation is called rectangular windowing because it is equivalent to multiplying a very long cosine wave with a shorter rectangular pulse.

DSPlay provides many types of windows all of which are found in the "Window" menu. Connect a 0.125Hz cosine block to a Rect window block. Then connect a Zero pad block to the output of the Rect window. Again set the input frame length to 128 points and set the output buffer length to 512 points.

a) Look at a 512 point FFT of the output of the Zero pad block. Is this the same as the output of the FFT block in problem 3(a)?

b) How wide, in Hertz, is one of the main lobes of the response?

c) While in plot mode, press the <c> key to get a magnitude and phase plot and then press the <d> key to convert the vertical axis of the magnitude plot to "dB". What is the drop (in dB) of the peak of the first sidelobe (the lobes immediately to the left and right of the main lobe)?

d) Now use a Hanning window instead of the Rectangular Window. Look at the output of the Hanning window block to see how this window affects the cosine. Again use the <d> key to get a dB plot of the magnitude of the output of the FFT block. What is the width of the main lobe? What is the drop (in dB) of the peak of the first sidelobe?

e) Replace the Hanning window with a Hamming window. Look at the shape of the cosine wave after traveling through the Hamming window block. Now look at the output of the FFT. What is the width of the main lobe? What is the drop (in dB) of the peak of the *highest* sidelobe?

Problem 6: *Circular Convolution*

As discussed in previous labs, the Convolve block does not actually compute the convolution sum. Instead, it computes the DFT (using the FFT) of the first input, then computes the DFT of the second input, multiplies the two DFTs together, and finally takes the inverse DFT to find the result.

This is circular convolution. Although it is faster than computing the convolution sum, circular convolution does have some drawbacks if the engineer is not careful.

Note: Since the output of the inverse DFT is always a complex signal, the output of the convolver block is also a complex signal. This is the reason that the convolver block produces a complex output even when both inputs are real.

Let $y(n)$ be defined as follows:

$$y(n) = \begin{cases} 1, & \text{if } n = 1, 2, 3, 4, 5, \textit{ or } 6 \\ 0, & \textit{otherwise} \end{cases} \tag{14}$$

and let $x(n)$ be defined as in problem 2.

a) Find and print or sketch the 8 point circular convolution: $y(n) * x(n)$.

b) Which points of the sketch for part (a) are affected by overlap caused by circular convolution?

c) Now find and print or sketch the 16 point circular convolution: $y(n) * x(n)$.

d) Compute a 128 point circular convolution of $x(n)$ and a 0.05Hz sine wave. Which points of the output (if any) are affected by overlap?

Problem 7: *More Circular Convolution*

Compute the 16 point convolution of $x(n)$ and $y(n)$ as in problem 6 but without using the Convolve Block. Instead, use the FFT and Inverse FFT blocks.

a) Print or sketch your flowgram.

b) Is the output the same as in problem 6(c)?

Laboratory Exercise 8

IIR Filter Design

Purpose:

This lab demonstrates both the impulse-invariant transformation and the bilinear transformation for IIR filter design from continuous-time filters. It makes use of the filter design tools that are part of DSPlay, thus providing practical experience with computer aided filter design tools.

Background Material:

The student should be familiar with IIR filter design from analog filters using impulse-invariance and the bilinear transformation. This material is covered in:

Oppenheim and Schafer [1988]: Section 7.1

Oppenheim and Schafer [1975]: Sections 5.0 - 5.2

Jackson: Chapter 8

Proakis and Manolakis: Sections 8.2 & 8.3

Strum and Kirk: Chapter 10

Roberts and Mullis: Sections 6.6 & 6.7 and Appendix 6A & 6B

DeFatta, Lucas, and Hodgkiss: Chapter 4

Kuc: Section 8.4

Rabiner and Gold: Chapter 4

Problem 1: *The Impulse Invariant Transformation*

The small filter design package included in DSPlay only computes IIR filters using the bilinear transform. As a result of this, we will have to use the IIR block to enter filters that we have designed on our own using the impulse invariant transform.

a) Assuming that the sampling period is $T = 1$, design a second-order discrete-time LPF by the impulse-invariant transformation with the cutoff frequency $\omega_c = 2\pi(0.1Hz)$. Enter your design into an IIR block in DSPlay. Plot the frequency response of your design. What is the drop in dB of the magnitude response at ω_c? Why is this drop not equal to 3 dB, as it is in the corresponding continuous-time filter? (**Hint:** the "Buffer mode" within the Plot mode will be useful.)

b) What is the drop (in dB) of the magnitude frequency response $|X'(\omega)|$ at $\omega = \pi/T$?

c) Now change your design so that $\omega_c = 2\pi(0.25Hz)$. What is the drop in dB of the magnitude response at ω_c?

d) What is the drop (in dB) of $|X'(\omega)|$ at $\omega = \pi/T$?

e) Which has more aliasing, the design of part (a) or the design of part (c)?

f) Change your design again so that $\omega_c = 2\pi(0.4Hz)$. Look at the frequency response of your new filter. Is this still a low-pass filter?

g) Is the impulse invariant method suitable to the design of high-pass filters? Justify your answer.

Problem 2: *The Bilinear Transformation: Butterworth Filters*

To access the filter design capability of DSPlay, press F2 and choose "Filter design". You now have a choice of types of filters to design. The "Butterworth IIR" and "Chebyshev IIR" filters are designed using the bilinear transformation. If you select a filter type, a query box appears and asks you for the specifications of the filter. Once these have been entered, the filter coefficients are computed (it may take a while). They will appear on the screen. You can browse through them using the arrow keys to move up and down. To exit the coefficients browser, press the escape key. The response of your new filter may be plotted using the "compute response" option in the menu. The horizontal axis of the response runs from $\omega = 0$ to (almost) $\omega = \pi/T$.

To use the filter as a block in a flowgram, do the following: 1) Save the coefficients of the filter in a file (this is a menu option). 2) Exit the filter design program by pressing the <Esc> key until the flowgram is back on the screen. 3) Create an empty block where the new filter is to be placed. 4) Position the "flashing box" cursor over the empty block. 5) Press "l" for load and "p" for parameters (these are also menu options). You will be asked to specify the filename where the filter coefficients are stored.

This newly created filter block behaves exactly like the IIR block that was used in earlier labs.

a) Use the filter design program in DSPlay to design a two-pole Butterworth low-pass filter with $\omega_c = 2\pi(0.4Hz)$. What is the drop in dB of the magnitude response at ω_c? Why is this different from problem 1?

b) What is the actual value of $|X'(\omega)|$ at $\omega = \pi/T$? What should it be and why?

   Note: to find the value of the filter at $\omega = \pi/T$, you will have to save the filter, exit the filter design program, and then load the filter into a suitable flowgram.

c) Using a Butterworth filter, what is the minimum number of poles needed to have a drop-off of at least 14 dB at $\omega = 2\pi(0.42Hz)$ when $\omega_c = 2\pi(0.40Hz)$?

d) What is the minimum number of poles required to have a drop-off of at least 14 dB at $\omega = 2\pi(0.22Hz)$ when $\omega_c = 2\pi(0.20Hz)$?

e) Why is the answer to part (c) different than the answer to part (d) when the bandwidth of the transition band is the same ?

f) DSPlay automatically does the spectral transformations necessary to create high-pass, band-pass, and notch (bandstop) filters. If you have some extra time, explore the responses of these filters.

Problem 3: *The Bilinear Transformation: Chebyshev Filters*

a) Use DSPlay to design a 5th order Chebyshev filter with $\omega_c = 2\pi(0.20Hz)$ and with a passband ripple of 1 dB. What type (I or II) of Chebyshev filter does DSPlay use for its design?

b) How many peaks (high and low points) are there in the ripple of the design of part (a)? How many peaks are there in the ripple of a 10th order filter? What is the relationship between the number of poles and the number of ripple peaks?

c) What is the minimum number of poles required to have a drop-off of at least 14 dB at $\omega = 2\pi(0.22Hz)$ when $\omega_c = 2\pi(0.20Hz)$ and the passband ripple is held to 1 dB ? Compare your answer to problem 2(d).

d) As with the Butterworth filter, DSPlay can convert Chebyshev low-pass filters into high-pass, band-pass, and band-stop filters. Why not spend some time to examine the responses of these filters?

Laboratory Exercise 9

FIR Filter Design via Windowing

Purpose:

This lab explores the design of FIR filters using windows. Designs include low-pass and high-pass filters using rectangular and Hamming windows. Linear phase is demonstrated by showing how symmetry of an input pulse results in symmetry of the output pulse.

Background Material:

FIR filter design using windows is covered in:

Oppenheim and Schafer [1988]: Sections 7.4 - 7.5

Oppenheim and Schafer [1975]: Sections 5.4 & 5.5

Jackson: Sections 9.0 - 9.1

Proakis and Manolakis: Section 8.1

Strum and Kirk: Sections 9.0 - 9.2

Roberts and Mullis: Sections 6.0 - 6.3 & 6.8

DeFatta, Lucas, and Hodgkiss: Section 5.3

Kuc: Section 9.2

Rabiner and Gold: Sections 3.1 - 3.16

Problem 1: *The Rectangular Window*

The most obvious way to create a Finite Impulse Response filter from an Infinite Impulse Response filter is to simply truncate the impulse response of the IIR filter. This truncation is equivalent to simply multiplying the infinite impulse response with a rectangular pulse (window). Multiplication in the time domain is equivalent to convolution in the frequency domain. Thus, the frequency response $H_d'(\omega)$ of the infinite impulse response of the desired filter $h_d(n)$ is convolved in frequency with the frequency response $W_r'(\omega)$ of the rectangular window $w_r(n)$. So to understand the effects of the truncation of $h_d(n)$, we must first know the frequency response $W_r'(\omega)$ of the window which performs the truncation.

Use the waveform or pulse block to generate a rectangular window. Then feed this window into an FFT block to find its frequency response.

a) Using N = 20, where N is the width of the rectangular window, find the drop in dB at the peak of the first side lobe of $W_r'(\omega)$. Also find the position (in Hz) of the peak of the first side lobe.

b) Using N = 40, repeat part (a).

c) Using N = 100, repeat part (a).

d) Looking at your results for parts (a) through (c): Is the dB drop at the first side lobe a function of N? Is the position of the peak of the first side lobe a function of N?

Problem 2: *Low Pass FIR Design Using Windowing*

DSPlay has a special utility program that computes the coefficients and response of windowed FIR filters. This utility program may be reached from DSPlay's block mode by pressing the F2 key, selecting "Filter design", and then selecting "Window FIR".

The Window FIR program operates much as the IIR Design programs operated in lab 8. The program asks for the type of window and then requests several parameters of the filter. After these parameters are entered, the program computes the coefficients for the filter requested. The frequency plot for the design may also be computed while in the design program.

After the filter is designed, it may be used by saving the coefficients to a file, escaping from the filter design program, and loading the file as the parameters of an empty block.

a) Design a filter using a rectangular window and the following specifications:

Low-pass filter with linear phase	
sample frequency:	$\omega_s = (1Hz)2\pi$
pass band edge:	$(0.2Hz)2\pi$
pass band gain:	1.0
stop band edge:	$(0.28Hz)2\pi$ through $(0.5Hz)2\pi$
stop band attenuation:	$18dB$

These specifications do not correspond exactly to the specifications in DSPlay, so they may need to be explained. At the pass band edge, we require no more than 3dB loss. In other words, the gain of the filter at this and lower frequencies should be 0.707 or greater[8]. The stop band edge specifies the frequency above which all frequencies must be attenuated by at least the stop band attenuation. An attenuation of 18 dB implies that the magnitude gain of the filter anywhere in the stopband should be less than 0.1259.

a) What is the minimum number of taps required to meet the specifications above?

What value did you use for the "cutoff" parameter for this minimum tap filter? Notice that in DSPlay, unfortunately, the "cutoff" parameter does not correspond to the -3 dB frequency.

b) The engineer who gave the specifications for part (a) has changed her mind. The filter is now to have a Stopband Attenuation of at least 50 dB. Using DSPlay, design a filter using the Hamming window that has the new desired stopband attenuation and meets the other specifications of part (a).

---

[8] Note, when the "d" key is used to create a dB plot in DSPlay; the program chooses the highest value of the plot for its zero dB point. Since there is significant ripple and overshoot in the passband, the dB plot will give the value of all points relative to the top of the overshoot. It may therefore be easier to use the linear plot.

What is the minimum number of taps required to meet these specifications?

What value did you use for the "cutoff" parameter for this minimum tap filter?

c) Use the design techniques given in your text to estimate the number of taps for an FIR filter with the specifications of part (b) using a Hamming window. Is this estimate higher or lower than the actual value obtained in part (b)?

Problem 3: *High-Pass and Band-Pass Windowed FIR filters*

DSPlay can design high-pass, band-pass, and stop-band windowed FIR filters; but it does this by only using low-pass filters!

To see how this is done, suppose you are given an ideal low-pass filter with impulse response $h_{LP}(n)$ and frequency response $H'_{LP}(\omega)$, shown in figure 17a. The high-pass frequency response can be written in terms of the low-pass frequency response as follows:

$$H'_{HP}(\omega) = 1 - H'_{LP}(\omega) \tag{15}$$

so the impulse responses are related by

$$h_{HP}(n) = \delta(n) - h_{LP}(n)\ . \tag{16}$$

The frequency response in figure 17a is real-valued, so the resulting filter

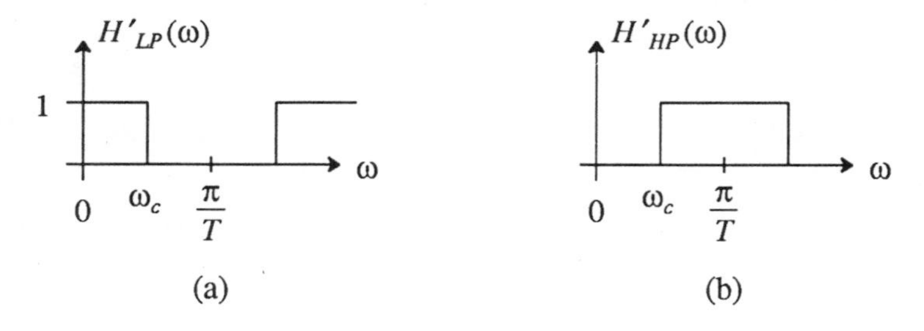

Figure 17. The frequency responses of an ideal low-pass and an ideal high-pass filter.

has a symmetric impulse response. Suppose (foolishly) that this filter can be realized with an FIR filter with $M+1$ coefficients, where M is an even number. This filter will be non-causal, but a causal filter with the same magnitude frequency response (and a linear phase term) has an impulse response

$$g_{LP}(n) = h_{LP}(n - M/2) . \tag{17}$$

Consequently, if you are given a causal FIR low-pass filter $g_{LP}(n)$ with linear phase, you would expect to be able to design a causal FIR high-pass filter as follows:

$$g_{HP}(n) = \delta(n-M/2) - g_{LP}(n) . \tag{18}$$

a) Use a rectangular window to find a 9 tap (M = 8) FIR low-pass filter with the "cutoff" parameter set to 0.25 Hz. Copy down or print (if the printers are working) the coefficients.

b) Now use a rectangular window to find a 9 tap (M = 8) FIR high-pass filter with a "cutoff" of 0.25 Hz. Also copy down or print these coefficients.

c) Do the high-pass tap values fit the formula (18)?

Problem 4: *Linear Phase Revisited*

The FIR filters of this lab all have linear phase. This means that all of the frequency components of the input are delayed by the same amount in the output. The IIR filters of the previous lab, however, had non-linear phase response. This problem explores the difference between the two types of filters.

a) Filter a periodic square pulse train (period = 80, pulse width = 10) with a low-pass, 20 tap, 0.05Hz cutoff, Hamming windowed FIR filter. Are the output pulses symmetric? What is the width of the resulting output pulses?

b) Now filter the same periodic square pulse train through a 6 pole, 0.05Hz cutoff, low-pass Butterworth IIR filter. Are the output pulses symmetric? Do the output pulses appear to be longer or shorter than those of part (a)?

Laboratory Exercise 10

Frequency Sampling and Equiripple FIR

Purpose:

The frequency sampling method for the easy design of filters with almost any desired frequency response is demonstrated. Also, the Parks-McClellan algorithm is used to generate multi-band optimal equiripple filters.

Background Material:

Addition information on FIR filter design using frequency sampling and equiripple methods may be found in the following:

Oppenheim and Schafer [1988]: Sections 7.6 - 7.7

Oppenheim and Schafer [1975]: Sections 5.6 & 5.7

Jackson: Sections 9.2 & 9.3

Proakis and Manolakis: Section 8.1

Strum and Kirk: Sections 9.4 - 9.6

Roberts and Mullis: Sections 6.4 & 6.5

DeFatta, Lucas, and Hodgkiss: Section 5.6.

Kuc: Section 9.5

Rabiner and Gold: Sections 3.17 and 3.23 - 3.40

Frequency Sampling: Example Procedures

The frequency sampling method is the most conceptually straightforward of all filter design techniques. To use it, specify the desired frequency response as completely as possible, and then compute the impulse response by computing the inverse DFT. The frequency-domain specification consists of equally spaced samples of the desired frequency response. Although this is simple in principle, we encounter some complications when trying to implement this technique using DSPlay.

In principle, the "Waveform" block can produce the samples of the desired frequency response, and its output can be fed into an Inverse FFT block to compute the impulse response of the filter. The number of filter coefficients (the order of the filter) will equal the number of samples given in the frequency domain by the waveform block.

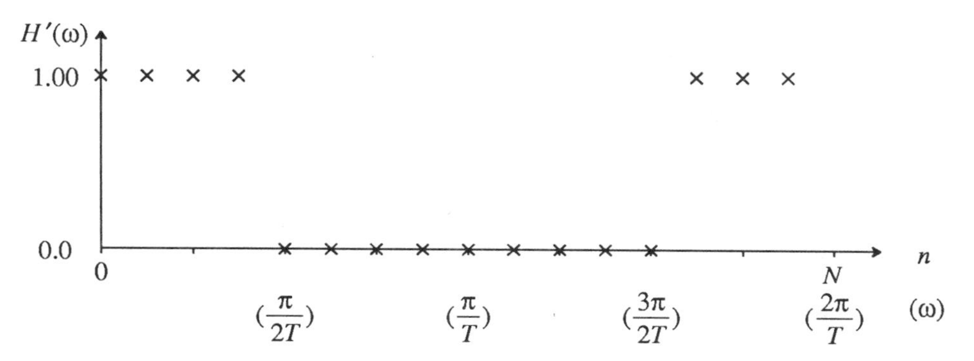

Figure 18. An example of the output of the Waveform block when a low-pass filter is desired. The indices n of the samples range from zero to $N-1$ and represent frequencies ranging from zero to almost $2\pi/T$. These frequencies are shown in parenthesis.

An example of the output of the Waveform block when a low-pass filter is desired is shown in figure 18. To get a real-valued FIR filter, the frequency samples must be symmetric about the point $\omega = \pi/T$, as shown. The cutoff frequency is somewhere between $3\pi/8T$ and $4\pi/8T$. The Waveform output of figure 18 can be generated using the following parameters: *cyclic*; (0,1); (3,1); (4,0); (12,0); (13,1); (16,1). Build this waveform block and connect it to an "Inv FFT" block with a buffer length of 16. The Inv FFT block is in the Spectrum menu. To check to see if the output is a reasonable impulse response you could connect it to an FFT block and look at the frequency response. If you use an FFT block with a buffer length of 16, you will get back exactly the output of the waveform block (check it if you doubt it), and if you use any other buffer length you will get a system error. Using a block length of 16 does not give you enough information about the frequency response, so a Zero pad block (in the Data menu) should be inserted between the Inv FFT block and the FFT block to increase the buffer length. Now examine the frequency response of the filter.

Although it appears that we have created a low-pass filter (albeit a rather poor one), there are some serious problems that we have ignored. Look at the output of the Inv FFT block, the signal we propose to use as the impulse response of the filter. It is symmetric, which implies that we have designed a linear phase filter, but its largest coefficients are at either end! Surely this is not what we intended. This is a consequence of the choice of samples produced in DSPlay by the Inverse FFT block. In principle, the inverse DFT is periodic, and DSPlay only needs to supply the samples in one period. However, as with any periodic signal, there is a choice of where to start.

Normally, we would use the waveform block to produce a set of real-valued samples with which we intend to represent the desired frequency response. Since these samples are real, the filter should be symmetric about $n = 0$. Thus, its impulse response would ideally be $h(-M/2)$ through $h(+M/2)$, where $h(n)$ is the output of the inverse FFT block. We could then create a causal filter by just shifting the samples by $M/2$ in time.

Unfortunately, DSPlay only handles causal signals. The inverse FFT block produces $h(0)$ through $h(M)$, not $h(-M/2)$ through $h(+M/2)$. To correct for this problem, we take advantage of the periodicity of an inverse DFT. Because $h(n)$ is in principle periodic with period N (even though DSPlay only gives us one period), the samples around $h(0)$ are the same as the samples around $h(0+N)$. Due to the causal nature of

DSPlay's signals, we cannot have the samples to the left of $h(0)$, but we can find the samples on both sides of $h(0+N)$. This can be accomplished in DSPlay with the system in figure 19. In this example, the Waveform block produces the desired frequency points (16 of them), the "Inv FFT" block converts the 16 frequency points into 16 time domain points, the Append and Cut blocks capture the points around $h(N)$ to create a causal filter, and the Zero pad and FFT blocks generate a detailed frequency response of the filter. Each of these blocks is explained in further detail below.

The append block takes any number of input frames an "appends" them together to make one larger output frame. For this to work in our system, the Waveform block must be made cyclic. To do this, be sure the "cyclic" parameter has value "Y", and make sure that the last point of your waveform has the time value of NT. This is important because the waveform block uses the time value of its last point to determine the period.

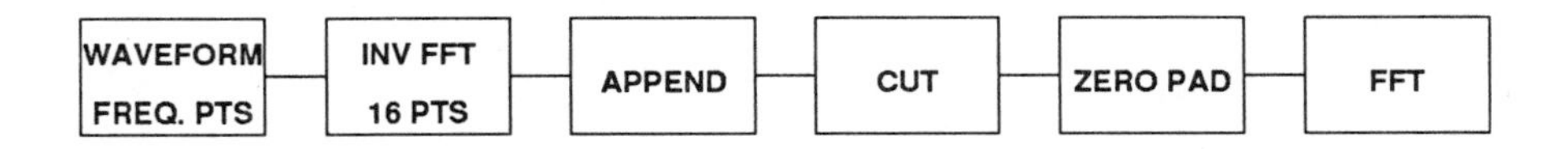

Figure 19. Flowgram for developing FIR filters using the frequency sampling method.

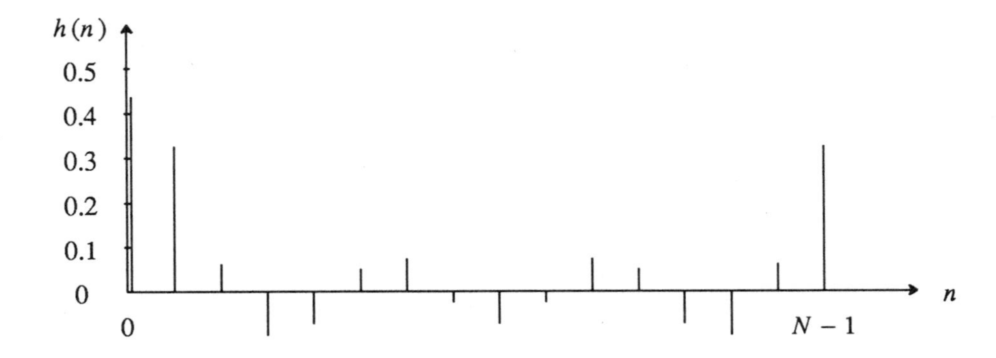

Figure 20. Output of the Inv FFT block.

The Inv FFT block converts the frequency points of the Waveform block into the time domain points $h(0)$ through $h(N-1)$ (see figure 20). The append block combines two output frames from the Inv FFT block into one frame. Use the append block (in the Data menu) to generate a plot like that in figure 21. Now the points centered around $h(N)$ are present. We can use the cut block to extract them.

The Cut Block produces as its output a subset of the input buffer. Use the Cut block (in the Data menu) to generate a plot like that in figure 22. Its parameters should be

input frame length: 32

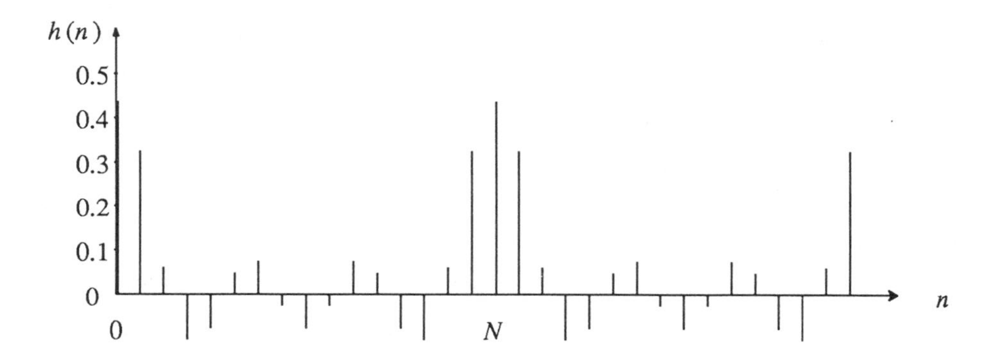

Figure 21. Output of the Append block.

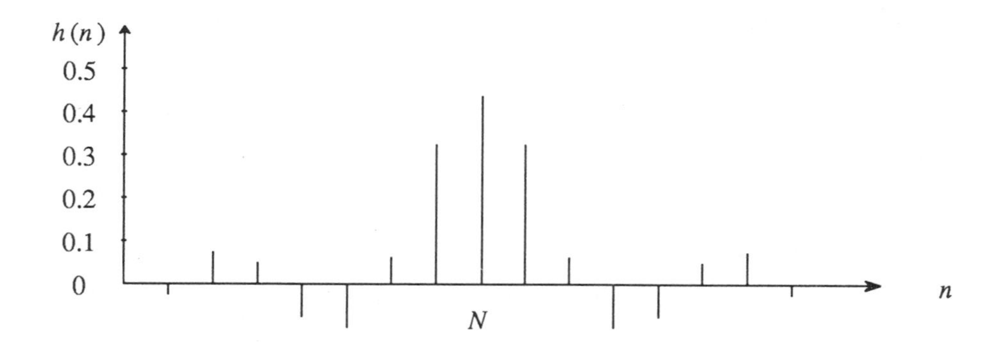

Figure 22. Output of the Cut block.

output buffer length: 15
input frame offset: 9

Now the output of the Cut block is the set of 15 points centered around $h(N)$. These are the taps of the FIR filter. In order to get a symmetric impulse response, The cut block should produce an odd number of samples (N-1). One significant sample gets discarded, and the resulting frequency response will be slightly different from the original samples.

Now use the Zero pad block to increase the buffer size for the FFT block. The input buffer length should be $N - 1 = 15$ and the output buffer length should match the length of the FFT corresponding to the desired resolution. Make these connections and examine the frequency response of the resulting filter. It should be a much better low-pass filter than what we got at first, as shown in figure 23 (generated with a 512 point FFT).

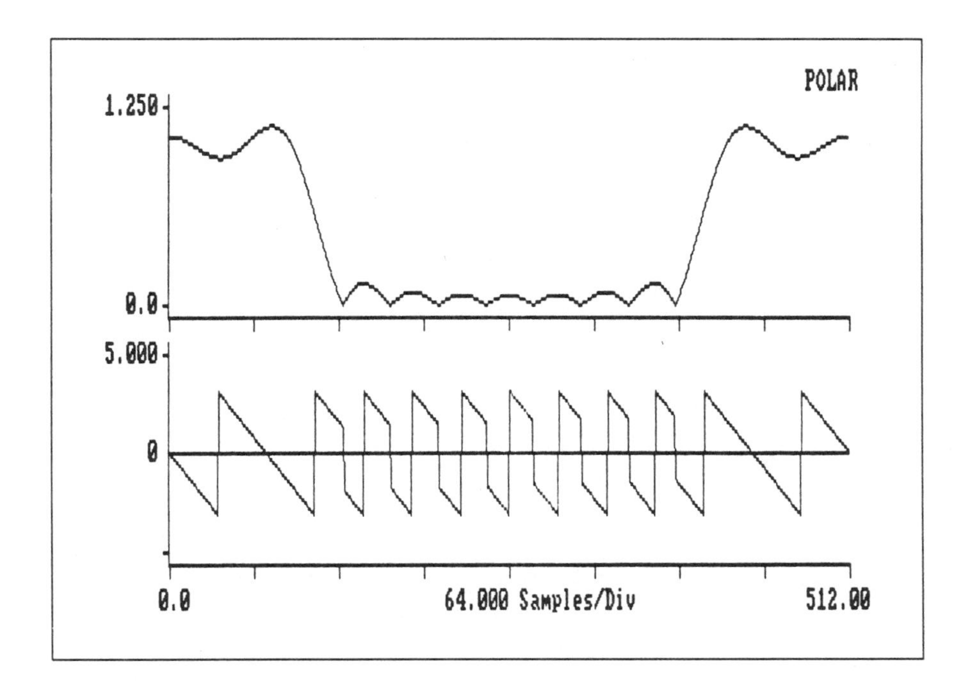

Figure 23. Output of the FFT block.

Problem 1: *Low Pass FIR Filter Design by Frequency Sampling*

a) For the system described above in the example procedure, what is the cutoff (3 dB) frequency of the FIR filter? Note that DSPlay is unable to determine the correct frequencies associated with the FFT. Thus, the horizontal axis of figure 23 is labeled only as samples. The horizontal axis actually runs from 0 to $2\pi/T$.

b) What is the drop (in dB) of the peak of the first sidelobe of the filter?

c) What is the width of the transition region? For this lab, the transition region is defined as the length (in Hz) between the cut-off frequency and the first zero of the stop band?

Problem 2: *Design with Transition Samples*

To increase the stopband attenuation, a transition sample is often used. This sample is somewhere between the passband and the stopband. For example, a transition sample of 0.2 is used in figure 24.

A filter is desired with the following characteristics:

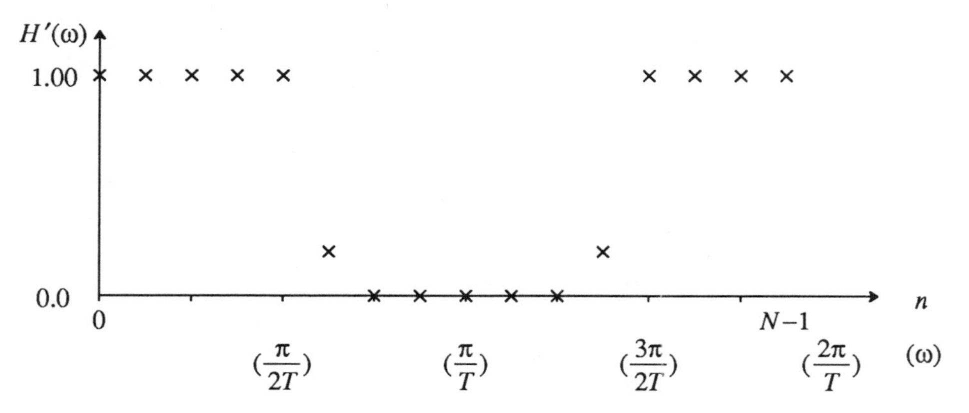

Figure 24. Frequency samples for a low-pass filter with a single transition sample.

Low-pass filter with linear phase	
sample frequency:	$\omega_s = (1Hz)2\pi$
pass band edge (-3dB point):	$(0.2Hz)2\pi$
pass band gain:	1.0
stop band edge:	$(0.30Hz)2\pi$
stop band attenuation:	$30dB$

a) Design this filter using the frequency sampling method and N = 16. Use a transition sample of 0.3904. Find the actual frequency response of your new filter. Does it meet the specifications above?

b) What is the minimum stopband attenuation (the attenuation at the peak of the first side lobe)?

c) What is the width of the transition region of this filter?

d) Now change your design so that the transition sample is 0.50. What is the minimum stopband attenuation?

e) How is the stopband in part (d) different from the stopband of part (a)?

Problem 3: *Oddly Shaped Filters*

The frequency sampling method can be used to generate more than just low-pass, high-pass, or band-pass filters. Any desired frequency response can be approximated using the frequency sampling method.

Use the frequency sampling method to produce a real FIR filter that has the frequency response shown in figure 25. Remember to sample over the entire frequency range: 0 to $2\pi/T$. Use $N = 32$.

a) What are the values used for your waveform block?

b) Where is the actual frequency response most different from the desired frequency response?

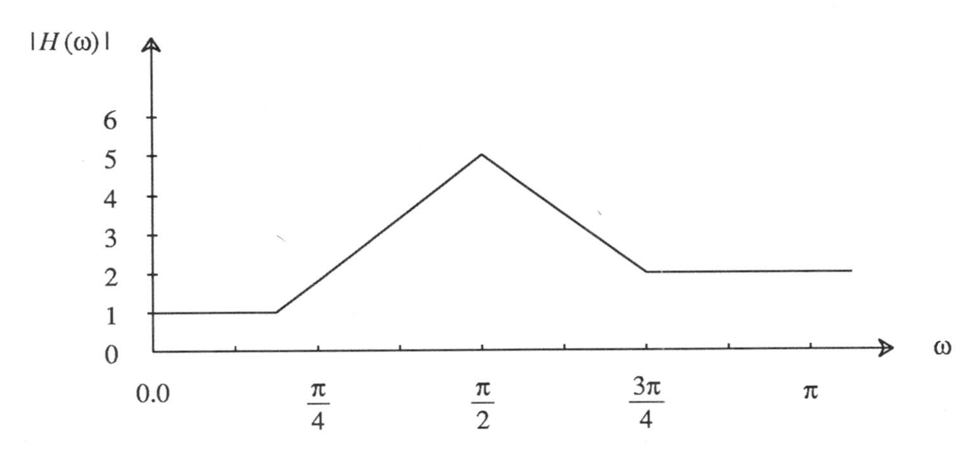

Figure 25. Desired frequency response for problem 3.

Equiripple Designs:

Problem 4: *Equiripple Filtering in DSPlay*

The equiripple filter design package in DSPlay uses the Parks-McClellan method for computing an FIR filter. Enter the Parks-McClellan FIR filter design program. The program offers a choice of three filter types. We will be using the "Multi pass/stop band" option exclusively for this lab so it should be selected.

The parameter request is quite formidable at first glance, but basically it is pretty simple. Each band can be a stopband, a passband or something in between. The band type is determined by the band gain. A band gain of 0 corresponds to a stopband and a band gain of 1 corresponds to a passband. The weight associated with each band is used to find the ratio of the ripple in different bands. After the parameters have been entered, the filter design program then generates the filter that has the smallest amount of ripple for the given order of the filter.

A few definitions are in order to make the measurement of ripple easier: δ_1 is the difference between the peak of the passband ripple and the desired value of the passband. δ_2 is the difference between the peak of the stopband ripple and the desired value of the stopband. K is the ratio of δ_1 to δ_2, $K = \delta_1/\delta_2$.

As an example, we want a 20 tap filter that has one tenth the ripple in the stopband compared to the passband. The filter should have a passband from 0 to $(0.2Hz)2\pi$ and a stopband from $(0.3Hz)2\pi$ to $(0.5Hz)2\pi$.

Our parameters would be as follows:

Sample Frequency: 1
Number of Taps: 20
Number of bands: 2
band 1) lo edge: 0.0, high edge: 0.2, band gain: 1, weight: 1
band 2) lo edge: 0.3, high edge: 0.5, band gain: 0, weight: 10

Try this filter out in DSPlay. Look at the stopband in the frequency domain. At first it appears flat because the ripple is small. If you zoom in, however, the ripple is apparent.

a) What is the value of δ_1 for this filter?

b) What is the value of δ_2 for this filter?

c) What is the value of K for this filter?

d) Is this filter extra-ripple?

e) Now redesign the filter using the same parameters but only 10 taps. What are the new values of δ_1, δ_2, and K?

Laboratory Exercise 11

Quantization Effects

Purpose:

This lab investigates the effects of coefficient quantization.

Background Material:

More information about coefficient quantization may be found in:

Oppenheim and Schafer [1988]: Sections 6.7 - 6.10

Oppenheim and Schafer [1975]: Sections 9.0 - 9.4

Jackson: Sections 9.0 - 9.1

Proakis and Manolakis: Sections 8.5 & 10.1 - 10.3

Strum and Kirk: Problem 2.23

Roberts and Mullis: Section 9.16 & 10.3

DeFatta, Lucas, and Hodgkiss: Section 2.5.4 & 9.4, 9.5

Kuc: Chapter 10

Rabiner and Gold: Sections 5.23 - 5.31

Coefficient Quantization:

Often, when implementing a digital filter, the coefficients may be specified with only limited precision. This causes movement of the actual positions of the poles and zeros of the filter and, thus, changes the frequency characteristics of the filter.

Problem 1: *IIR Filter Coefficient Quantization*

Use the filter design program to create a 2 pole low-pass Butterworth filter with $\omega_c = (0.2Hz)2\pi$.

Copy down the coefficients to your new filter. The first row of coefficients corresponds to A_0, A_1, and A_2. The second row of coefficients corresponds to B_1, and B_2, where the IIR filter is defined by:

$$H(z) = \frac{A_0 + A_1 z^{-1} + A_2 z^{-2}}{1 + B_1 z^{-1} + B_2 z^{-2}} \tag{19}$$

a) Quantize your coefficients to 3 bits with a bit pattern as follows: $a_{\Delta}bc$ Where a is the sign bit, Δ is the binary point, and b and c are two bits after the binary point. Note that (using two's complement notation) this quantization only gives coefficients with the following values: -1.0, -.75, -.5, -.25, 0, .25, .5, and .75

   Create a small table of the five original coefficients along with your new quantized coefficients.

b) Implement your new quantized coefficients in an IIR block. Where is the new ω_c ? How is the stopband different from the stopband of the non-quantized filter response?

c) Now perform the quantization again using one more bit in the fraction. That is, use numbers of the form: $a_{\Delta}bcd$ Note that (using two's complement notation) the coefficients are limited to the following values: -1.0, -.875, -.75, -.625, ... , .5, .625, .75, and .875 .

   Create another small table of your new, 4 bit quantized coefficients.

d) Implement these new coefficients in an IIR filter block. Where is ω_c for this filter? Is the stopband better, or worse, than the filter of part (b)?

Problem 2: *Narrow Band Low Pass IIR Coefficient Quantization*

Narrow-band low-pass and high-pass filters are especially vulnerable to coefficient quantization when implemented in direct form.

Design a 2 pole, low-pass, Butterworth filter with $\omega_c = (.1Hz)2\pi$. Copy down these coefficients. Note that the range of coefficient values here is greater than the filter of problem 1.

a) Quantize these coefficients using 6 bits. Note that one of the coefficients is greater than one. Thus we will use our bits as follows: $ab_{\Delta}cdef$ where a is the sign bit and Δ is the binary point. With our bits defined in this manner and with two's complement notation, the quantized coefficients can have the following values: -2, -1.9375, -1.875, -1.8125, -1.750, -1.6875, ... , -.0625, 0 , .0625, .1250, .1875, ..., 1.8125, 1.875, and 1.9375 .

   Make a table of the values of your quantized coefficients.

b) Create an IIR filter block with the quantized coefficients of part (a) and find the frequency response of your filter. What is the new value of ω_c?

Problem 3: *Very Narrow Band Low Pass IIR Filters*

When a very narrow band low-pass filter is required, normal filter design techniques don't work when the coefficients are quantized.

a) Design a 2 pole, low-pass, Butterworth filter with $\omega_c = (0.03Hz)2\pi$. Copy down these coefficients. Now quantize these coefficients to 6 bits using the same bit pattern as in problem 2 above.

   What is wrong with the new, quantized filter?

b) One way to get around this problem is to lower the sampling rate, perform the low-pass filtering, and then to increase the sampling rate back to its original level. Down-sampling and up-sampling are supported in DSPlay via the Decimate and Interpolate blocks which may both be found in the Data category.

   Implement a low-pass filter with $\omega_c \approx (0.03Hz)2\pi$. Use only the filter from problem 2(b) (you may need more than one such filter) and the decimate and interpolate blocks.

Printout (or sketch) the flowgram of your filter system. On your printout (or sketch) indicate the anti-aliasing filter block. Also indicate the values of the parameters of the interpolate and decimate blocks.

Problem 4: *Coefficient Quantization of an FIR filter*

Use the filter design program to design a high-pass FIR filter with $\omega_c = (0.3Hz)2\pi$. Use a Hanning window for your design and use 51 taps. Save your filter and then load it into an empty block.

a) Find the frequency response of your new filter. What is the drop (in dB) at the peak of the first side lobe? What is the drop (in dB) at DC?

b) Now quantize all 51 taps. Don't panic!! There is an easy way to do this in DSPlay. Connect the impulse response of the FIR filter into a Quantize Block[9] and connect the Quantize block to an FFT to see the frequency response of the newly quantized FIR filter. Use a range of -1 to +1 and 12 bits to quantize your filter coefficients.

   What is the new drop (in dB) at the peak of the first side lobe? What is the drop (in dB) at DC? Does this filter have linear phase in the pass band?

c) Now change the quantize block so that the quantization is done to 10 bits. Find the frequency response of your 10 bit taps.

   What is the drop (in dB) at the peak of the first side lope? What is the drop at DC?

d) Finally, set the quantizer to 8 bits and find the new frequency response.

   What is the minimum drop (in dB) in the stopband (between 0 and 2.5Hz) ? Does this filter still have linear phase?

e) In general, will quantizing the coefficients affect the linear phase characteristic of an FIR filter? Why or why not?

---

[9] The Quantize block does the same operation to its input samples that you performed manually on the coefficients of problems 1 and 2. The limits on your ranges should be powers of two. If a number to be quantized is larger than a full-scale limit, it is set to that limit. For some unknown reason, the quantize block is located in the Arithmetic menu instead of the Non-Linear menu.

APPENDIX A
DSPLAY COMMAND SUMMARY

1. Editing Modes

Two basic editing modes are used throughout. The *flowgram mode* is indicated by the cursor being in the "home" position, at the bottom of the screen. The *block mode* is indicated by the cursor (a flashing box) being in the flowgram. In flowgram mode, commands apply to the entire flowgram. In block mode, commands apply to the block pointed to by the cursor.

2. A Note About Buffers

When you run a flowgram, DSPlay makes a decision about the number of samples to process. The default is 128. However, you can control this number by specifying a buffer length somewhere in the system that is different from 128. For example, the FFT and convolver blocks require that you specify a buffer length. All other blocks in the system must have compatible buffer lengths, or an error results.

DSPlay begins with signal sources and generates all of the data in the output buffers of the blocks. This data remains in the buffers until the next time the block is run. Blocks can be run one at a time, in which case they take their data from the buffers of their predecessors. Buffers can be displayed, saved to files, and restored from files.

If you are not using any block that has its buffer length as one of its parameters (for example the FFT or convolver), but you wish to control the buffer length nonetheless, then you can use the "append" block. By properly selecting its parameters, the append block does nothing except set the system buffer length.

3. Special Keys

The following table summarizes the action of several special function keys. The action often depends on the editing mode. The table also indicates the Lab in which the command is first used and described. If the command is not explicitly mentioned in any of the labs, an "N" is listed.

Special Keys — All Modes		Lab
<F1>	The basic help key (very useful).	1
<F2>	Various utilities, described below.	1
<F8>	Redraw the screen.	1
<PrtSc>	Print the screen image.	N
<Ctrl><Num Lock>	Stops scrolling of data Any key will resume data scrolling	N

Special Keys — Flowgram Mode		
<Esc>	Escape from current mode.	1
<up arrow>	Switch to block mode.	1

Special Keys — Block Mode		
<arrow>	(Four keys, left, right, up, down)	
	1) move the cursor,	1
	2) draw lines to make connections,	1
	3) select inputs of a block.	1
<Ins>	1) Insert a new block in the flowgram,	1
	2) Enter the line drawing mode.	1
<Del>	Delete a block or connection.	1
<Enter>	A shortcut for <E> followed by <P>. Use to specify parameters of a block.	1
<Ctrl><Enter>	Shortcut used to avoid entering a long list of zero-valued parameters.	4
<home>	Switch to flowgram mode.	1
<Alt><D>	Plot. Shortcut for <D> followed by <B>.	1

4. Basic Commands

The following table summarizes the basic DSPlay commands. Note that these commands need not be memorized since they are listed at the top of the screen. Most commands have a series of subcommands, indicated in italics. These subcommands are promted with a pull-down menu, and may be selected from the menu by moving the cursor, or by typing the one capitalized letter in the command. The tables below list that letter.

Several of these commands operate on "buffer data", which is the output of a block from the last run. The buffer data can be plotted, tabulated, saved, or loaded.

Commands — Flowgram Mode			Lab
L	Load a previously saved flowgram		N
S	Save current flowgram		N
R	Run the flowgram.		1
E	Edit the flowgram parameters: name, description, sample rate, display info.		1
D	Display.		N
	F	*Flowgram parameters.*	N
	S	*Display buffer data of marked blocks.*	N
P	Print various things.		N
Q	Quit DSPlay.		1

Commands — Block Mode			
L	Load disk files.		N
	B	*Load buffer data from a disk file.*	N
	P	*Load block parameters from a file.*	5
S	Save disk files.		N
	B	*Save buffer data to a disk file.*	N
	P	*Save block parameters to a file.*	N
R	Run part of the flowgram.		1
	B	*Block only.*	1
	U	*Until block (run all predecessors).*	1
E	Edit a block.		1
	P	*Parameters. Define the block function.*	1
	C	*Copy tagged block into current block.*	1
	M	*Move tagged block into current block.*	N
	T	*Tag a block for copying or moving.*	1
	U	*Untag. Remove the tag on a block.*	1
D	Display.		1
	B	*plot Buffer data.*	1
	T	*Tabulate buffer data.*	2
	L	*Landscape. Create a 3-D plot using segments of the buffer data.*	N
	P	*Parameters. Display the parameters of the current block.*	N
P	Print.		
	B	*print a plot of Buffer data.*	N
	T	*print a tabulation of Buffer data.*	N
	P	*print block parameters.*	N
Q	Quit DSPlay.		1
M	Mark a block for display.		1

5. Utilities

In addition to the main program, DSPlay includes a number of utility programs. These utilities may be accessed while in DSPlay by pressing the <F2> key. To leave a utility, press the <Esc> key until the utility has been exited. A short summary of the utilities is included in the table below. An longer summary of some of these utilities immediately follows the table.

Commands — Utility	
D	Directory. List the contents of a directory.
H	Change Directory. Change the default directory.
F	Filter Design.
	P *Parks-McCellan FIR design.*
	W *Window FIR design.*
	B *Butterworth IIR design.*
	C *Chebyshev IIR design.*
T	Text Editor
S	.asc → .dat Convert file to DSPlay data format.
A	.dat → .asc Convert file to ASCII data format.

5.1. Directory

This utility acts just like the DOS DIR command. If no directory is given, the utility list the contents of the default directory.

5.2. Change Directory

Allows the user to change the default directory. The default directory is displayed near the bottom of the screen.

5.3. Filter Design

This utility allows for the design of several types of filters. Once the parameters for the desired filter type have been entered, the program displays a listing of the coefficients of the filter. To exit this coefficient listing press the <Esc> key. This will cause a menu to appear. The menu allows the user to calculate and plot the filter response, replot the filter response, re-design the filter, or save the filter coefficients.

To use a designed filter in DSPlay, the following steps must be taken:

1) Design the filter using the desired filter type.

2) Press the <Esc> key to get out of the coefficient list and to the filter menu.

3) Save the filter coefficients.

4) Use the <Esc> key to leave the Filter Design utility and to return to the Flowgram.

5) Move the flashing box cursor over an emtpy block.

6) Press LP for Load Parameters to load in the saved filter. Use the same filename as in step 3 above.

5.4. .asc -> .dat

This utility converts files stored in ASCII format into the DSPlay binary data format. The ASCII file must start with the line ".DAT" . Also, The data must be in a column. For the exact format, generate a DSPlay data file and use the ".dat -> .asc" utility. Use the resulting ASCII file as a model.

5.5. .dat -> .asc

This utility converts binary data files created by DSPlay to ASCII format.

6. Block Types

The following tables summarize the heart of DSPlay, the block types. In block mode, typing EP or <Enter> over an empty block brings up a menu of block categories, listed below. A category may be selected by moving the cursor in the menu and typing <Enter>, or by typing the single key letter that is capitalized in the menu. A new menu appears listing the blocks in the category. The lab in which they are first described is listed in the last column. If the function is not specifically mentioned in any lab, it has an "N" in the last column.

KEY	DESCRIPTION	LAB
G	**subGram category**	N
I	**Inputs category (signal generators)**	
F	Get File. Retrieves data stored in a file	N
P	Pulse. Piecewise linear pulse (can be periodic)	1
S	Sinewave generator	2
C	Cosine generator	5
R	Random noise generator	N
G	Gaussian noise generator	N
D	DC signal generator	1
W	Waveform. Linearly connect points given	1
F	**Filter category**	
F	FIR. Finite impulse reponse filter.	4
I	IIR. Rational Z transform system function.	2
R	Cascaded IIR sections.	N
A	Adaptive filter.	N
C	Convolves two signals.	1
W	**Window category**	
H	Hamming window	7
N	Hanning window	7
T	Triangle window	N
C	Custom window	N
R	Rectangular window	7

S	Spectrum category	
F	FFT	3
I	Inverse FFT	7
D	DFT	7
N	iNverse DFT	N
P	Power spectrum	N
C	Correlation	N

C	Control category	
D	Delay	4
L	Limiter	N
P	Peak	N

A	Arithmetic functions category	
A	Add two signals	6
S	Subtract two signals	6
M	Multiply two signals	1
D	Divide	N
W	Weight. Multiply a signal by a constant	N
V	Absolute Value of each sample	N
C	Accumulate a total or average of the samples	N
Q	Quantize each sample	11

T	Trigonmetric functions category	
S	Sine of the input signal	N
C	Cosine of the input signal	N
T	Tangent of the input signal	N
I	Arcsine of the input	N
O	Arccosine of the input	N
A	Arctangent of the input	N

N	Non-linear functions (and some linear)	
G	Log base 10 of the input	N
L	Ln. Natural log.	1
E	Exponential (e^x).	1
S	Square	N
Q	Square Root	N
P	Power. Raise samples to a given power	N
R	Root.	N
I	Integral. Discrete-time version of an integrator	1
D	Derivative.	N

D	Data category	
R	Rect → Polar, for complex data	N
P	Polar → Rect, for complex data	N
I	Interpolate between samples	11
D	Decimate, downsample	11
A	Append buffers together	10
C	Cut out part of a buffer	10
Z	Zero Pad a signal	4
O	Complex signals → Real signals	6
E	Real signals → Complex signals	6
S	Section a large buffer into smaller buffers	N
J	Conjugate complex data	N
V	Reverses the input signal	4

O	Output category	
F	Put File. Writes the input signal to a file	N

6.1. SubGrams

A subgram is a block representing a flowgram. Subgrams must be used if the flowgram is larger than the screen, and can always be used to improve the readability of the flowgram. To enter the subgram screen, use the <Pg Dn> key. This screen may be exited via the <Pg Up> key. The screen associated with the subgram can be thought of as a magnification of the subgram block.

The output of the right hand, center block (already on the screen) will be the output of the subgram. The subgram may have 0, 1, or 2 inputs. The vertical line at the left edge of the screen corresponds to the primary input.[10] Thus, any line that is connected to the vertical line on the left edge of the screen will also be connected to the center input notch of the subgram block. The vertical line just to the right of the left edge corresponds to the secondary input.

Subgrams, like flowgrams, may be saved and loaded. A subgram may be run like any other block. Subgrams may be nested as deeply as desired, until memory is exhausted.

6.2. Inputs

Get File:

The output of the Get File block is the data contained in the file specified in the parameters. When the Get File block is run, it starts at the beginning of the file (or at the Input File Offset, if specified). If the Get File block is connected to an Append or Accumulate block, it will read continuously from the file.

The easiest way to use the Get File block is to ignore the Input Offset and Frame Overlap parameters. Make sure when using the Get File block that the Data Type parameter matches the type of data that is actually in the file.

Buffer Note: This block uses a fixed buffer size. Any other blocks connected to the Get File block that have a fixed buffer (ie. an FFT block) must have the same buffer length as the Get File block.

Pulse:

The Pulse block creates a pulse train as its output. This block can be used to produce a unit impulse, or square, triangular, and saw-toothed pulses.

The parameters of the pulse block (pulse width, rise time, and fall time) are all assumped by the program to be in *seconds*. Thus, the number of *samples* per pulse will depend upon the sampling frequency set in the flowgram parameter menu.

---

[10] The primary input is always the input notch in the center of the block. When line drawing mode is entered, the cursor always appears over the primary input. The secondary input is the input notch just below (or sometimes above) the center notch.

Note: Even though a non-cyclic pulse does not have a period, the program requires one anyway. It is usually best to chose a large number like 1000.

Sinewave Generator:

This block generates a sine wave with the frequency, amplitude, and phase shift specified in the parameters.

Cosine Generator:

This block generates a cosine wave.

Random Noise Generator:

This block generates white noise with a uniform distribution.

Gaussian Noise Generator:

This block generates gaussian random noise with a mean and variance determined by the parameters.

DC Signal Generator:

The output of this block is constant; a DC signal.

Waveform:

The Waveform block generates its output by connecting the points given as parameters with straight lines. The last point given will be the value of the output of the block after the time value of the last point.

Note: For cyclic waveforms, the period of the waveform will be equal to the time value of the last point of the waveform parameter set. (See lab 10).

This is a difficult block to use. It is a good general practice to check the output of a Waveform block to make sure it is correct.

6.3. Filters

FIR:

The FIR block runs the input signal through the Finite Impulse Response filter specified in the parameters. The form of the filter is

$$y(n) = A_0 x(n) + A_1 x(n-1) + A_2 x(n-2) + \cdots + A_{128} x(n-128)$$

where $y(n)$ is the output signal and $x(n)$ is the input signal.

When done entering the desired coefficients, hold down the <Enter> key or press <Ctrl><Enter> to skip over undesired coefficients.

IIR:

The Infinite Inpulse Response filter block filters the input signal with an IIR filter. The filter is implemented using direct form II. The coefficients are found directly from the transfer function of the filter:

$$H(z) = \frac{A_0 + A_1 z^{-1} + A_2 z^{-2} + \cdots + A_{20} z^{-20}}{1 + B_1 z^{-1} + B_2 z^{-2} + \cdots + B_{20} z^{-20}} \quad (20)$$

Remember to make sure that the filter is in the correct form before entering coefficients.

Cascaded IIR Sections:

This block consists of ten second-order IIR filters cascaded together. Each row represents a different IIR filter. The numerator coefficients of each filter are the first three parameters in each row. The denominator coefficients are the last two parameters in each row.

Each section (row) of the cascade is an IIR filter and has the same representation and parameter set as the IIR filter block above.

Adaptive Filter:

The adaptive filter block is not used in this lab.

Convolver:

The Convolver block computes the circular convolution of its two input signals. Because this is circular convolution and not linear convolution, extra care must be taken when using this block (see lab 7).

Note: The output of this block is always complex, even if both of the input signals were real. This is a result of the inverse FFT that is used at the end of the circular convolution.

Buffer Note: This block has a fixed buffer. As a result of this, the buffer size parameter of this block must match the buffer size of any other fixed buffer blocks it is connected to (directly or indirectly). Also, the buffer length must be a power of two.

6.4. Windows

Hamming Window:

This block multiplies the input signal by a Hamming window. The Hamming window used will be just as long as the block size, presumably the length of the input signal. That is, if the input signal is 128 points long, a 128 sample Hamming window is applied.

Hanning Window:

This blocks multiplies the input signal by the Hanning window.

Triangular Window:

This window is also know as the Bartlett window.

Custom Window:

The Custom window works very much like the Waveform block. The window is defined by connecting the points give as parameters with straight lines. The last point of the custom window is used until the end of the input signal.

Rectangular Window:

This block multiplies the input signal by a rectangular window. Note that since the width of the window is the same as the width of the input signal, this block actually does nothing at all!

6.5. Spectrum

FFT:

This block computes the Fast Fourier Transfrom of the input signal. The output of this block is always complex.

Buffer Note: This block uses a fixed buffer length. All other blocks that are connected to it (directly or indirectly) must have compatible buffer lengths. Otherwise, a "Buffer Size Error" will result when the flowgram is run. Also, the buffer size for this block must be a power of two.

Inverse FFT:

This block computes the inverse FFT of the input signal. The input to the inverse FFT block must be a complex signal. The output of this block will always be a complex signal.

Buffer Note: This block uses a fixed buffer length. All other blocks that are connected to it (directly or indirectly) must have compatible buffer lengths. Also, like the FFT, the buffer size for this block must be a power of two.

DFT:

This block computes the DFT of the input signal. This block can be used to compute the DFT at selected frequencies or over the entire range of frequencies. The output of this block is always a complex signal.

The Low Limit and High Limit parameters are used by DSPlay to determine the range of frequencies to calculate. The full frequency range would be setting the Low Limit to 0 and the High Limit to (Buffer size) - 1. For example, a full range DFT on a buffer size of 128 would have a Low Limit = 0 and a High Limit = 127.

Buffer Note: This block uses a fixed buffer length. All other blocks that are connected to it (directly or indirectly) must have compatible buffer lengths. Also, like the FFT, the buffer size for this block must be a power of two.

Inverse DFT: The Inverse DFT block computes the inverse DFT over a specified frequency range. This block requires its input to be complex. The output of this block is also a complex signal.

Buffer Note: This block uses a fixed buffer length. All other blocks that are connected to it (directly or indirectly) must have compatible buffer lengths. Also, like the FFT, the buffer size for this block must be a power of two.

Power Spectrum:

This block computes the power spectrum of the input signal.

Buffer Note: This block uses a fixed buffer length. All other blocks that are connected to it (directly or indirectly) must have compatible buffer lengths. Also, because this block uses an FFT, the buffer size must be a power of two.

Correlation:

This block computes the cross-correlation between two signals. The output of this block is always a complex signal. To compute the auto-correlation of a signal, simply connect both inputs of the Correlation block to the same input signal.

Buffer Note: This block uses a fixed buffer length. All other blocks that are connected to it (directly or indirectly) must have compatible buffer lengths.

6.6. Controls

Delay:

This block delays the input signal by the number of samples given as the Buffer Delay parameter. The output signal will have leading zeros corresponding to the length of the delay.

Limiter:

This block is used to limit the positive and negative extremes of the input signal. All values that exceed the limit will be set to the limit. The Data Type parameter must be set to the data type of the input signal.

Peak:

This block finds the peak value or peak location of the input signal. The output signal is a single point. To look at this single point the Display -

Tabulate buffer command must be used while in flowgram mode.

Buffer Note: This block has a transitional buffer. All fixed buffer blocks connected to the input of the Peak block must have the same buffer length as the Peak block. All fixed buffer blocks connected to the output of the Peak block must have their buffer sizes set to 1.

6.7. Arithmetic

Add:

The Add block adds the two input signals point by point. The two input signals may be both real or both complex. However, if one input is complex and the other is real, an error will result.

Subtract:

This block subtracts the secondary input from the primary input on a point by point basis. The primary input is always the input notch in the center of the block. When line drawing mode is entered, the cursor always appears over the primary input. The secondary input is the input notch just below (or sometimes above) the center notch.

The type of both inputs must be the same. DSPlay allows two real or two complex signals to be subtracted. A mixture of the two types will result in an error.

Multiply:

This block multiplies the two input signals point by point. The two input signals must both be of the same type or an error will result.

The multiply block may also be used to square a signal. This may be accomplished by connecting both inputs to the same signal.

Divide:

The Divide block divides the primary input by the secondary input. The primary input is always the input notch in the center of the block. When line drawing mode is entered, the cursor always appears over the primary input. The secondary input is the input notch just below (or sometimes above) the center notch.

The two input signals must be of the same type or an error will result. Also, the secondary input signal must not have any zero valued samples or an error will result from dividing by zero.

Weight:

This block multiplies every point in the input signal by the Gain parameter. It then adds to every point in the input block the value of the Offset parameter.

The Data Type parameter must match the data type of the input signal. If the input signal is real, the imaginary parts of the Gain and Offset parameters are not used.

Absolute Value:

This block computes the absolute value of each of the input samples.

Accumulate:

The input to this block is run again and again (as many times as the number of frames specified as a parameter). It then totals or averages these runs.

For example, if the number of frames were set to 4 and a total was desired, the Accumulate block would run its input 4 times, creating four frames. The first sample of the output would be (first sample of the first run of the input) + (the first sample of the second run of the input) + (the first sample of the third run) + (the first sample of the fourth run of the input). The other samples of the output are computed in the same way.

The average is computed by dividing each sample of an accumulated total by the number of frames.

Quantize:

This block quantizes the input signal to the levels specified by the parameters. The number of levels is 2^n where n is the Resolution parameter.

6.8. Trigonometric functions

Sine:

This block calculates the sine of the input data. The input is in radians.

Cosine:

The output of this block is the cosine of the input. The cosine is calculated on a point by point basis. The input is in radians.

Tangent:

The output of this block is the tangent of the input signal. The input is in radians.

Arcsine:

This block computes the Arcsine of the input signal. All samples of the input signal must be between +1 and -1. The output of the Arcsine block will be in radians and between $-\pi/2$ and $\pi/2$.

Arccosine:

This block computes the Arccosine of the input signal. All samples of the input signal must be between +1 and -1. The output of the Arccosine block will be in radians and between 0 and π.

Arctangent:

This block computes the Arctangent of the input signal. The output of the Arctangent block will be in radians and between $-\pi/2$ and $\pi/2$.

6.9. Non-Linear (and some linear)

Log:

The Log block computes the log base 10 of the input signal, point by point. The input signal must be real. All samples of the input signal must be positive.

Ln:

The Ln block computes the log base e (natural log) of the input signal, point by point. The input signal must be real. All samples of the input signal must be positive.

Exponential:

This block raises e to the power of each input sample. The output of the block is (e^x) where x is an input sample.

Square: This block computes the square of the input signal on a point by point basis. The input signal must be real.

Square Root:

The output of this block is the positive square root of the input signal. The input signal must be real and positive.

Power:

This block raises each sample of the input signal to the power given by the Exponent parameter. The output is x^a where x is an input sample and a is the Exponent parameter.

Root:

This block takes the root (given as a parameter) of the input signal on a point by point basis.

Integral:

This block computes the sum of all previous input data points. This gives a running total of the input signal. This block operates on real data only.

Derivative:

This block estimates the slope of the input data at each point. The Derivative block computes this by averaging the slopes of the line segments connecting the current point to the points just before and just after the current point.

6.10. Data

Rectangular → Polar:

This block converts the complex input signal from rectangular coordinates to polar coordinates.

Note: All blocks in DSPlay that accept complex input signals expect those signals to be in rectangular coordinates.

Polar → Rectangular:

This block converts a complex signal that is in polar coordinates to rectangular coordinates.

Interpolate:

The Interpolate block adds an integer number of points between each point in the input signal. The number of points inserted between each point will be: (Interpolation Factor) - 1. The value of the inserted samples is computed by linearly interpolating between the input samples.

Buffer Note: This block has a transitional buffer. The number of data points in the input buffer may be different from the number of points in the output buffer.

The input buffer is a fixed buffer. All fixed buffer blocks connected to the input of the Interpolate block must have the same buffer size as the Input Frame Length parameter or an error will result.

The output buffer is also a fixed buffer. All fixed buffer blocks connected to the output buffer must have the same buffer length as the output buffer of the Interpolate block. The length of this output buffer is: (Interpolation Factor) × (Input Frame Length).

Decimate:

This block only puts every (Decimation Factor)th input sample into the output. If the Decimation Factor parameter is 1, the output will just be a copy of the input. If the Decimation Factor is 2, the output will simply be every 2nd input sample. If the Decimation Factor is 3, the output signal will consist of every 3rd input sample and will only be one third as long as the input signal.

Buffer Note: This block has a transitional buffer. The number of data points in the input buffer will be different from the number of points in the output buffer.

The input buffer is a fixed buffer. All fixed buffer blocks connected to the input of the Decimate block must have the same buffer size as the Input Frame Length parameter or an error will result.

The output buffer is also a fixed buffer. All fixed buffer blocks connected to the output buffer must have the same buffer length as the output buffer of the Decimate block. The length of this output buffer is: (Input Frame Length) divide by (Decimation Factor).

Append:

The Append block connects many smaller buffers together into one large buffer. The blocks connected to the input of the Append block are run the number of times specified by the Number of Frames to Append parameter. The input signals generated by each of these runs are then appended together to form one large buffer.

Buffer Note: Both the input and output buffers are fixed. Thus, any other fixed buffer blocks connected to the input or output of the Append block must have the same buffer length as the buffer of the Append block it is connected to.

The length of the input buffer of the Append block is just the Input Frame Length. The length of Append block's output buffer is: (Input Frame Length) $\times$ (Number of Frames to Append).

Cut:

This block cuts out a portion of the input signal to create the output signal. The output of this block is the Output Buffer Length number of points starting at the Input Frame Offset of the input signal. Input data before the Input Frame Offset and after the (Input Frame Offset) + (Output Buffer Length) is not transfered to the output.

Buffer Note: This block has a transitional buffer. The number of data points in the input buffer will be different from the number of points in the output buffer.

The input buffer is a fixed buffer. All fixed buffer blocks connected to the input of the Cut block must have the same buffer size as the Input Frame Length parameter or an error will result.

The output buffer is also a fixed buffer. All fixed buffer blocks connected to the output buffer must have the same buffer length as the output buffer of the Cut block.

Zero Pad:

This block pads the input signal with trailing zeros. The output of the block will simply be the input signal with zeros added on to the end. The Output Frame Length parameter must be greater than the Input Frame Length parameter.

Buffer Note: This block has a transitional buffer. The number of data points in the input buffer will be different from the number of points in the output buffer.

The input buffer is a fixed buffer. All fixed buffer blocks connected to the input of the Zero Pad block must have the same buffer size as the Input Frame Length parameter or an error will result.

The output buffer is also a fixed buffer. All fixed buffer blocks connected to the output buffer must have the same buffer length as the output buffer of the Zero Pad block.

Complex → Real:

This block converts a complex signal to a real signal. The real signal will consist of either the real part or the imaginary part of the complex input signal.

Real → Complex:

This block converts the type of the signal from real data to complex data. The real data in the input signal may appear as the real or imaginary part of the complex output signal. This is determined by the parameter. The part of the output not chosen will be set to zero. To form a complex signal with both real and imaginary parts being non-zero, two of these blocks plus and adder can be used.

Section:

This block produces as its output a section of the input buffer. In this respect, it operates very much as the Cut block does. However, if the Section block is connected to an Append or Accumulate block, the Section block will continue to break off sections of its input buffer and send them on to its output buffer.

Buffer Note: This block has a transitional buffer. The number of data points in the input buffer will be different from the number of points in the output buffer.

The input buffer is a fixed buffer. All fixed buffer blocks connected to the input of the Section block must have the same buffer size as the Input Frame Length parameter or an error will result.

The output buffer is also a fixed buffer. All fixed buffer blocks connected to the output buffer must have the same buffer length as the output buffer of the Section block.

Conjugate:

This block conjugates the input signal. The input to this block must be a complex signal. The output of this block is always a complex signal.

Reverse:

This block reverses the input signal.

6.11. Output

Put File

Stores the input signal unto a file on the disk. If the Number of Iterations parameter is not set to one, the flowgram connected to the input of the Put File block will be run repeatedly and the data from each of these runs will be appended unto the file on the disk. This file may be read back into a flowgram using the Get File block.

Buffer Note: This block has a transitional buffer. The number of data points in the input buffer may be different from the number of points in the output buffer.

The input buffer is a fixed buffer. All fixed buffer blocks connected to the input of the Put File block must have the same buffer size as the Input Frame Length parameter or an error will result.

The output buffer is also a fixed buffer. All fixed buffer blocks connected to the output buffer must have the same buffer length as the output buffer of the Put File block. The length of this output buffer is: (Input Frame Length) $\times$ (Number of Iterations).

APPENDIX B
DSPlay SOFTWARE INFORMATION

1. DSPlay (ED) Product Specification

Name	- DSPlay™ (ED) (Educational Version of DSPlay™)
Description	- Graphic DSP Programming Environment for the PC
Version	- V1.0 May - 1988
Environment	- For IBM®(PC, XT, AT) and Compatables
System Requirements	- 512K Memory minimum, 640K recommended - DOS 2.0 or higher - One 360K floppy disc drive - Monitor CGA or EGA
Media	- 5 1/4" 360K floppy disc
Documentation	- None supplied

2. Differences From Retail Product, DSPlay V1.2

DSPlay Vl.2 will support:

* 8087 and 80287 Coprocessors
* User defined blocks (Turbo Pascal®)
* Analog I/O using Burr-Brown's PCI20,000 series or RC Electronic's IS-16
* Buffers to 4095 points (DSPlay (ED) supports 512)

3. The DSPlay ™ Software

The *DSPlay*™ Package is a digital signal processing software package designed to transform the IBM®PC into an easy-to-use DSP workstation. The software allows the user to specify a DSP application in terms of block diagrams called FlowGrams™. FlowGrams are created using a graphic editor with pull-down lists and menu selections. Once the user has created their FlowGram the program can execute and display the results. Additional features include Filter Design programs and Text Editor.

DSPlay (ED) is a limited version of the standard *DSPlay* product and supplied with this book for use as a teaching aid. The limitations of *DSPlay* (ED) over the *DSPlay* Package are as follows:

1. The Math Co-processor (8087 or 80287) is not supported in *DSPlay* (ED).
2. No Analog input or Analog output Boards/Hardware are supported by *DSPlay* (ED).
3. The buffer size of any individual block is limited to 512 points in *DSPlay* (ED).
4. The user may not write and include their own Block Functions in *DSPlay* (ED).

4. Installing and Running DSPlay (ED)

There are 13 files supplied on the *DSPlay* (ED) diskette:

1.	DSPLAY	EXE	-	Main program
2.	FILDES	EXE	-	Filter Design overlay
3.	EGAFXRES	EXE	-	Print Screen driver for EGA
4.	DSPLOGO	EXE	-	Writes initial screen
5.	DSP_BIT	MAP	-	Data required by DSPLOGO.EXE
6.	FORM	LIB	-	DSPLAY.EXE initialization data
7.	CGA	BGI	-	Graphics Driver for CGA
8.	EGAVGA	BGI	-	Graphics Driver for EGA
9.	DSPLAY	HLP	-	TEXT to support HELP function
10.	XLAB5	NOD	-	Example
11.	SLAB6	NOD	-	Example
12.	LPLAB5	NOD	-	Example
13.	README			

To install *DSPlay* (ED) on your Fixed Disk:

1. Start DOS in the usual way.
2. Make a new DOS Directory named "DSPLAY".
3. Copy the files from the *DSPlay* (ED) distribution diskette into the new DSPLAY directory.

To start *DSPlay* (ED) from your Fixed Disk:

1. Execute the DOS Change Directory command make the DSPLAY directory your current default directory.
2. Initialize graphics support for printing as follows:
   a. If your PC has CGA, execute the DOS "GRAPHICS" command.
   b. If your PC has EGA, enter "EGAFXRES".
3. Enter "DSPLAY" to start the DSPlay (ED) program.

Note: *DSPlay* requires a copy of the DOS file COMMAND.COM in your root directory to support overlay activities.

Running from Diskette

If your PC has CGA:

1. Start DOS in the usual way with a system diskette in drive A.
2. Execute DOS "GRAPHICS" command to prepare your printer driver to support *DSPlay* (ED) Print commands.
3. Insert the *DSPlay* (ED) diskette in drive A.
4. Enter "DSPLAY" to start the *DSPlay* (ED) program.

If your PC has EGA:

1. Start DOS in the usual way with a system diskette in drive A.
2. Insert the *DSPlay* (ED) distribution diskette in drive A.
3. Enter "EGAFXRES" to install the supplied printer driver for EGA.
4. Enter "DSPLAY" to start the *DSPlay* (ED) program.

After *DSPlay* (ED) has initialized, you can remove the distribution diskette from drive A and insert another with more capacity for data. In addition to your data, that diskette should contain the following files:

1. COMMAND COM - copied from a DOS system diskette (to support overlays)
2. FILDES EXE
3. DSPLAY HLP

DSPlay™ is a trademark of Burr-Brown Corporation.
IBM®PC Is a registered trademark of International Business Machines.
FlowGram™ Is a trademark of Burr-Brown Corporation.

5. Burr-Brown Software License Agreement

BURR-BROWN CORPORATION
SOFTWARE LICENSE AGREEMENT

This is a legal document which is an agreement between you, the **Licensee**, and **Burr-Brown Corporation. By opening this sealed diskette package, Licensee agrees to become bound by the terms of this Agreement, which includes the Software License, Software Disclaimer of Warranty, and Hardware Limited Warranty** (collectively the "Agreement").

This Agreement constitutes the complete Agreement between Licensee and Burr-Brown Corporation. If Licensee does not agree to the terms of this Agreement, do not open the diskette package. Promptly return the unopened diskette package and the other items (including written materials, binders or other containers, and hardware, if any) that are part of this product to Burr-Brown Corporation for a full refund.

1. ***Grant of License and Software Ownership.*** In consideration of payment of the **License** fee, which is a part of the price **Licensee** paid for this product, **Burr-Brown**, as Licensor, grants to **Licensee**, a nonexclusive right to use and display this copy of a **Burr-Brown** software program (hereinafter the "**Software**") on a single computer. This **License** is not a sale of the original **Software** or of any copy. **Licensee** owns the magnetic or other physical media on which the **Software** is originally or subsequently recorded or fixed, but **Burr-Brown** retains title and ownership of the **Software** recorded on the original diskette copy(ies) and all subsequent copies of the **Software**, regardless of the form or media in or on which the original and other copies may exist. If the single computer on which **Licensee** uses the **Software** is a multiuser system, the **License** covers all users on that single system. **Burr-Brown** reserves all rights not expressly granted to **Licensee**.

2. ***Copy Restrictions.*** This **Software** and the accompanying written materials are copyrighted. Unauthorized copying of the **Software**, including **Software** that has been modified, merged, or included with other software, or of the written materials is expressly forbidden. **Licensee** will be held legally responsible for any copyright infringement that is caused or encouraged by its failure to abide by the terms of this License. Subject to these restrictions, and if the **Software** is not copy-protected, **Licensee** may make one (1) copy of the **Software solely for backup purposes. Licensee** shall reproduce and include the copyright notice on the backup copy.

3. ***Use Restrictions.*** **Licensee** may transfer the **Software** from one computer to another provided that the **Software** is used on only one computer at a time. **Licensee** may not distribute copies, including electronic transfer of copies, of the **Software** or accompanying written materials to others.

4. ***Authorized Modification and Integration.*** **Licensee** may, solely for its own purposes and use, modify, adapt and merge the **Software** with other software and electronically transfer copies of the modified, adapted, or merged **Software** provided that the modified, adapted, or merged **Software** is used on only one computer at a time. **Licensee** may not distribute copies, including electronic transfer of copies, of the modified, adapted or merged **Software** or accompanying written materials to others. All modifications, adaptations, and merged portions of the **Software** constitute the **Software** licensed to **Licensee**, and the terms and conditions of this Agreement apply to same.

5. ***Transfer Restrictions.*** This **Software** is licensed only to **Licensee** and may not be transferred to anyone without the prior written consent of **Burr-Brown**. Any authorized transferee of the **Software** shall be bound by the terms and conditions of this Agreement. In no event shall **Licensee** transfer, assign, rent, lease, sell, or otherwise dispose of the **Software** on a temporary or permanent basis except as expressly provided herein.

6. ***Termination.*** This **License** is effective until terminated. This **License** will terminate automatically without notice from **Burr-Brown** if **Licensee** fails to comply with any provision of this **License**. Upon termination **Licensee** shall destroy the written materials and all copies of the **Software**.

7. ***Update Policy.*** From time to time, **Burr-Brown** may create updated versions of the **software**. At its option, **Burr-Brown** may make such updates available to the **Licensee** and may correct software error(s) brought to its attention by **Licensee. Burr-Brown** may require the payment of an update fee and the return of a Registration Card.

8. ***Disclaimer of Warranty and Limited Warranty.*** **The Software and accompanying written materials (including instructions for use) are provided "as is" without warranty of any kind. Further, Burr-Brown does not warrant, guarantee, or make any representations regarding the use, or the results of the use, of the software or written materials in terms of correctness, accuracy, reliability, currentness, or otherwise. The entire risk as to the results and performance of the software is assumed by Licensee and not by Burr-Brown or its distributors, agents, or employees.**

Burr-Brown warrants to the original **Licensee** that (a) the diskette(s) on which the **Software** is recorded is free from defects in materials and workmanship under normal use and service for a period of ninety (90) days from the date of delivery as evidenced by a copy of the bill of sale and (b) the hardware accompanying the **Software** is free from defects in materials and workmanship under normal use and service for a period of one (1) year from the date of delivery as evidenced by a copy of the bill of sale. Further, **Burr-Brown** hereby limits the duration of any implied warranty(ies) on the diskette or such hardware to the respective periods stated above. Some jurisdictions do not allow limitations on duration of an implied warranty, so the above limitation may not apply to **Licensee**.

The above are the only warranties of any kind, either express or implied, including but not limited to the implied warranties of merchantability and fitness for a particular purpose, that are made by Burr-Brown on this Burr-Brown product. No oral or written information or advice given by Burr-Brown, its distributors, agents, or employees shall create a warranty or in any way increase the scope of this Warranty, and Licensee may not rely on any such information or advice. Burr-Brown's Warranty as herein set forth shall not be enlarged, diminished or affected by, and no obligation or liability shall arise or grow out of, Burr-Brown's rendering of technical advice, facilities or service in connection with this Burr-Brown Software. This Warranty gives Licensee specific legal rights; Licensee may have other rights, which vary from one jurisdiction to another.

Neither Burr-Brown nor anyone else who has been involved in the creation, production, sale or delivery of this product shall be liable for any direct, indirect, consequential, or incidental damages (including damages for loss of business profits, business interruption, loss of business information, and the like) arising out of the use of or the inability to use such product even if Burr-Brown has been advised of the possibility of such damages. Because some jurisdictions do not allow the exclusion or limitation of liability for consequential or incidental damages, the above limitation may not apply to Licensee.

9. ***Limitation of Remedies.*** **Burr-Brown's** entire liability and **Licensee's** exclusive remedy shall be:

(a) The replacement of any diskette(s) which does not meet **Burr-Brown's** Limited Warranty and which is returned to **Burr-Brown Corporation** with a copy of **Licensee's** bill of sale; or

(b) If **Burr-Brown Corporation** is unable to deliver a replacement diskette(s) which is free of defects in materials and workmanship, **Licensee** may terminate this Agreement by returning the diskette(s) together with accompanying written materials, binders or other containers, and hardware (if any) and **Licensee's** money will be refunded.

Any replacement diskette or hardware will be warranted for the remainder of the original warranty period or thirty (30) days, whichever is longer. If failure of the diskette or hardware has resulted from accident, abuse, or misapplication, **Burr-Brown** shall have no responsibility to replace the diskette or hardware or refund the purchase price.

10. ***U.S. Government Restricted Rights.*** The **Software** and documentation is provided with **restricted rights**. Use, duplication, or disclosure by the Government is subject to restrictions as set forth in subdivision (b) (3) (ii) of The Rights in Technical Data and Computer Software clause at 252.227-7013. Contractor/manufacturer is **Burr-Brown Corporation, P.O. Box 11400, Tucson, Arizona, 85734 USA.**

11. ***Miscellaneous.*** This agreement is governed by the laws of the State of Arizona.

If **Licensee** has any questions concerning this Agreement, **Licensee** may contact **Burr-Brown** in writing: **Burr-Brown Corporation, P.O. Box 11400, Tucson, Arizona, 85734 USA.**

Fill in the card on the reverse side to request more information on Burr-Browns' DSP Products,

or

To order the full *DSPlay*™ Language Package at the special price of $395.

(This offer is made only to original puchasers of this book. Please use the original card; no duplicates acceptable.)

POSTAGE WILL BE PAID BY ADDRESSEE

Burr-Brown Corporation
ATTN: DSP Support, Quest
P.O. Box 11400
Tucson, AZ 85775-3403